Learning Pandas 2.0

A Comprehensive Guide to Data Manipulation and Analysis for Data Scientists and Machine Learning Professionals

Matthew Rosch

Published by: GitforGits
Publisher: Sonal Dhandre
www.gitforgits.com
support@gitforgits.com

Printed in India

First Printing: April 2023

ISBN: 978-8119177066

Cover Design by: Kitten Publishing

Content

Preface

"Learning Pandas 2.0" is an essential guide for anyone looking to harness the power of Python's premier data manipulation library. With this comprehensive resource, you will not only master core Pandas 2.0 concepts, but also learn how to employ its advanced features to perform efficient data manipulation and analysis.

Throughout the book, you will acquire a deep understanding of Pandas 2.0's data structures, indexing, and selection techniques. Gain expertise in loading, storing, and cleaning data from various file formats and sources, ensuring data integrity and consistency. As you progress, you will delve into advanced data transformation, merging, and aggregation methods to extract meaningful insights and generate insightful reports.

"Learning Pandas 2.0" also covers specialized data processing needs, such as time series data, DateTime operations, and geospatial analysis. Furthermore, this book demonstrates how to integrate Pandas 2.0 with machine learning libraries like Scikit-learn, TensorFlow, and PyTorch for predictive analytics. This will empower you to build powerful data-driven models to solve complex problems and enhance your decision-making capabilities.

What sets "Learning Pandas 2.0" apart from other books is its focus on numerous practical examples, allowing you to apply your newly acquired skills to tricky scenarios. By the end of this book, you will have the confidence and knowledge needed to perform efficient and robust data analysis using Pandas 2.0, setting you on the path to become a data analysis powerhouse.

In this book you will learn how to:

- Master core Pandas 2.0 concepts, including data structures, indexing, and selection for efficient data manipulation.
- Load, store, and clean data from various file formats and sources, ensuring data integrity and consistency.
- Perform advanced data transformation, merging, and aggregation techniques for insightful analysis and reporting.
- Harness time series data, DateTime operations, and geospatial analysis for specialized data processing needs.
- Visualize data effectively using Seaborn, Plotly, and advanced geospatial visualization tools.
- Integrate Pandas 2.0 with machine learning libraries like Scikit-learn, TensorFlow, and PyTorch for predictive analytics.

GitforGits

Prerequisites

Whether you're a seasoned data professional or just starting your journey in data science, "Learning Pandas 2.0" is the perfect resource to help you harness the power of this cutting-edge library. This book is an absolute resource of practical implementation of Pandas 2.0 in every possible data manipulation and analysis project.

Codes Usage

Are you in need of some helpful code examples to assist you in your programming and documentation? Look no further! Our book offers a wealth of supplemental material, including code examples and exercises.

Not only is this book here to aid you in getting your job done, but you have our permission to use the example code in your programs and documentation. However, please note that if you are reproducing a significant portion of the code, we do require you to contact us for permission.

But don't worry, using several chunks of code from this book in your program or answering a question by citing our book and quoting example code does not require permission. But if you do choose to give credit, an attribution typically includes the title, author, publisher, and ISBN. For example, "Learning Pandas 2.0 by Matthew Rosch".

If you are unsure whether your intended use of the code examples falls under fair use or the permissions outlined above, please do not hesitate to reach out to us at support@gitforgits.com.

We are happy to assist and clarify any concerns.

Acknowledgement

I owe a tremendous debt of gratitude to GitforGits, for their unflagging enthusiasm and wise counsel throughout the entire process of writing this book. Their knowledge and careful editing helped make sure the piece was useful for people of all reading levels and comprehension skills. In addition, I'd like to thank everyone involved in the publishing process for their efforts in making this book a reality. Their efforts, from copyediting to advertising, made the project what it is today.

Finally, I'd like to express my gratitude to everyone who has shown me unconditional love and encouragement throughout my life. Their support was crucial to the completion of this book. I appreciate your help with this endeavour and your continued interest in my career.

CHAPTER 1: INTRODUCTION TO PANDAS 2.0

Understand Pandas

Pandas is an open-source Python library that provides data manipulation, analysis, and cleaning tools designed to make data analysis fast, easy, and intuitive. It has gained immense popularity in the data science and analytics community due to its high-performance, user-friendly, and easy-to-understand capabilities. Pandas stands for "Python Data Analysis Library," and its primary data structure, the DataFrame, is modeled after the R programming language's data frame. Pandas 2.0 is the latest version of the library and offers significant performance improvements and new features over its predecessor.

Data manipulation and analysis are crucial tasks in data science, and Pandas excels at providing efficient and versatile tools to perform these tasks. Data manipulation includes tasks like data cleaning, data transformation, and data aggregation. Data analysis, on the other hand, encompasses tasks such as data exploration, visualization, and statistical analysis.

Introducing Pandas 2.0

Pandas is built on top of the NumPy library, which is an essential library for scientific computing with Python. This dependency allows Pandas to leverage the power and efficiency of NumPy arrays while providing additional features tailored for data analysis. The two main data structures that Pandas provides are Series and DataFrame, both of which are built on top of NumPy arrays.

A Series is a one-dimensional, labeled array capable of holding any data type. It is similar to a Python list or a NumPy array but has an index that provides labels for each element. The index can be integer-based, like the default Python list, or it can be user-defined with custom labels.

A DataFrame, on the other hand, is a two-dimensional, size-mutable, and heterogeneous tabular data structure with labeled axes (rows and columns). It is essentially a collection of Series objects, with each column being a Series. DataFrames are designed to handle a wide variety of data types, including numerical, categorical, and text data, making it an incredibly versatile tool for data manipulation and analysis.

Why Pandas for Data Manipulation & Analysis?

Pandas provides numerous functionalities that make it an essential library for data manipulation and analysis:

Data ingestion: Pandas can read data from various file formats like CSV, Excel, JSON, and

SQL databases, making it easy to load and parse data from various sources.

Data cleaning: Pandas provides tools for handling missing data, such as filling missing values with default values or interpolating values based on surrounding data. It also enables users to filter, rename, and drop columns or rows, making it easy to clean and preprocess data.

Data transformation: Pandas offers powerful tools for reshaping and pivoting DataFrames, making it easy to transform data into the desired format. It also provides functionality for concatenating, merging, and joining DataFrames, allowing users to combine and restructure datasets.

Data aggregation and summarization: Pandas supports numerous aggregation functions like sum, mean, median, and standard deviation, as well as custom aggregation functions. GroupBy functionality allows users to perform these aggregations on specific subsets of data based on one or more columns.

Time series functionality: Pandas provides comprehensive tools for working with time series data, including resampling, rolling windows, and shifting data forward or backward in time. This makes it especially useful for analyzing time-dependent data, such as financial or sensor data.

Data exploration: Pandas provides convenient methods for filtering, sorting, and selecting specific subsets of data. This makes it easy to explore and understand datasets.

Data visualization: Pandas integrates seamlessly with popular data visualization libraries like Matplotlib, Seaborn, and Plotly, enabling users to create a variety of visualizations to better understand their data.

Performance: Pandas is designed for high performance and can handle large datasets efficiently. It leverages vectorized operations and Cython-optimized functions to provide fast computation times, even for complex data manipulation tasks. The library also supports parallel and out-of-core computations, making it suitable for large-scale data processing.

Compatibility: Pandas is compatible with a wide range of other Python libraries and tools, such as scikit-learn, TensorFlow, and PySpark, enabling seamless integration into a data science workflow. This makes it an excellent choice for building end-to-end data processing pipelines.

Community and documentation: Pandas has a large and active community that contributes to its development and maintenance. This ensures that the library stays up-to-date with the latest advancements in data analysis techniques. Furthermore, Pandas has extensive documentation and numerous tutorials, making it easy for newcomers to learn and understand its capabilities.

To sum it up, Pandas is a powerful and versatile Python library that plays a significant role in data manipulation and analysis. Its efficient and user-friendly tools make it a go-to choice for data scientists and analysts. By providing seamless data ingestion, cleaning, transformation, aggregation, summarization, exploration, and visualization capabilities, Pandas makes the entire data analysis process faster, easier, and more intuitive. Additionally, its high-performance features, compatibility with other Python libraries, and active community support make Pandas an essential tool for any data scientist or analyst looking to leverage the power of Python for their data analysis tasks.

Install and Configure Pandas 2.0

To install Pandas 2.0 and configure it for use in your Python environment, follow these step-by-step instructions. Note that these steps assume you have Python already installed on your system.

Install pip

Pip is the package installer for Python. It is used to install and manage Python packages, including Pandas. If you do not have pip installed, follow these instructions:

For Windows and macOS

Download the get-pip.py file from https://bootstrap.pypa.io/get-pip.py

Open a terminal (Command Prompt on Windows or Terminal on macOS) and navigate to the directory where you downloaded the file.

Run the following command:

```
python get-pip.py
```

For Linux

Open a terminal and run the following command:

```
sudo apt-get install python3-pip
```

Create Virtual Environment

Creating a virtual environment allows you to manage dependencies separately for each of your projects. This prevents conflicts between different packages and their dependencies.

Open a terminal (Command Prompt on Windows or Terminal on macOS/Linux).

Navigate to your project directory or the directory where you want to create the virtual environment.

Run the following command to create a new virtual environment named "env":

```
python -m venv env
```

Activate the Virtual Environment

On Windows: env\Scripts\activate

On macOS/Linux: source env/bin/activate

Install Pandas 2.0

With the virtual environment activated, you can now install Pandas 2.0 using pip:

In the terminal, run the following command:

```
pip install pandas==2.0
```

Verify the Installation

To verify that Pandas 2.0 has been installed successfully, perform the following steps:

Start the Python interpreter by running the python command in your terminal.

Import Pandas with the following command: import pandas as pd

Check the Pandas version by running: print(pd.__version__)

If the output shows "2.0", the installation was successful.

Configure Pandas

Pandas has sensible default settings for most use cases. However, you may want to configure specific options to suit your needs. To configure Pandas options, you can use the pd.set_option() function.

Following is an example of how to set the maximum number of rows displayed when printing a DataFrame:

Start the Python interpreter by running the python command in your terminal.

Import Pandas with the following command: import pandas as pd

Set the desired option, for example: pd.set_option("display.max_rows", 500)

You can find more options in the Pandas documentation:
https://pandas.pydata.org/pandas-docs/stable/user_guide/options.html

Now that Pandas 2.0 is installed and configured, you can start using it for data manipulation and analysis in your Python projects. Remember to import Pandas at the beginning of your Python scripts using import pandas as pd.

Pandas Version History

Pandas has a rich history of development, with numerous versions released over the years. Each version has introduced new features, performance improvements, and bug fixes. Below is a brief overview of the major version history of Pandas up to 1.3.x, followed by a detailed description of Pandas 2.0.

Pandas 0.1.x (January 2011)

The initial release of Pandas provided the foundation for DataFrame and Series data structures, along with basic I/O operations, time series functionality, and basic data manipulation capabilities.

Pandas 0.2.x to 0.9.x (2011-2012)

These versions introduced many new features, such as MultiIndex, pivot tables, and additional I/O support for various file formats like Excel and HDF5. The library underwent substantial improvements in performance, API stability, and documentation.

Pandas 0.10.x to 0.23.x (2013-2018)

During this period, Pandas continued to expand with new functionalities, including support for categorical data, window functions, and enhanced support for missing data. There were ongoing performance optimizations and refinements to the existing APIs.

Pandas 0.24.x to 1.0.x (2019-2020)

This phase saw the addition of the extension array interface, nullable integer and boolean data types, and improved support for custom data types. Pandas 1.0 marked a major milestone in terms of API stability and maturity, deprecating older and less-used features while refining the library's core functionality.

Pandas 1.1.x to 1.3.x (2020-2021)

These releases brought more performance improvements, additional I/O options, new string data types, and improved compatibility with other Python libraries like Dask and CuPy. The library's documentation and testing infrastructure also saw significant enhancements.

Pandas 2.0 (2023)

As a major release, Pandas 2.0 brings significant changes to the library, focusing on improving performance, addressing long-standing issues, and providing better support for modern data analysis requirements.

Some key features and improvements in Pandas 2.0 include:

Performance improvements

Pandas 2.0 features substantial performance improvements, particularly for large datasets. The library leverages advancements in array computing, modern hardware capabilities, and parallel processing to deliver faster and more efficient data manipulation and analysis operations.

Enhanced type system

The type system in Pandas 2.0 has been refined to provide better support for nullable data types, user-defined data types, and extension arrays. This allows users to work with a wider range of data types and customize Pandas to better suit their specific needs.

New APIs and functionality

Pandas 2.0 introduces several new APIs and features that make it easier to manipulate and analyze data. This includes new methods for working with hierarchical data, advanced window functions, and improved support for working with geospatial and time series data.

Improved interoperability

Pandas 2.0 aims to provide better interoperability with other Python libraries and tools, making it easier to integrate Pandas into a data science and analytics workflow. This includes enhanced compatibility with libraries like Dask for parallel and distributed computing, as well as better integration with visualization libraries like Matplotlib, Seaborn, and Plotly.

Simplified and more consistent API

With Pandas 2.0, the library's API has been simplified and made more consistent. This makes it easier for users to learn and work with Pandas, and it reduces the chances of encountering unexpected behavior.

Modernized codebase

The Pandas 2.0 codebase has been modernized and refactored to improve maintainability, readability, and extensibility. This includes removing deprecated features and code, adopting modern Python language features, and embracing best practices for software development. As a result, Pandas 2.0 is more robust and easier to contribute to for the community of developers.

Enhanced documentation and tutorials

The documentation and learning resources for Pandas 2.0 have been expanded and improved, making it easier for users to learn and understand the library's features and capabilities. The updated documentation includes more examples, clearer explanations, and a more intuitive organization of content.

Improved error messages and debugging

Pandas 2.0 provides more informative error messages and better debugging support, helping users quickly identify and resolve issues in their code. This includes more descriptive error messages, better tracebacks, and improved support for using debuggers like pdb.

New deprecation policy

With the release of Pandas 2.0, the library has adopted a more structured deprecation policy, providing users with a clear roadmap for future changes and a more predictable upgrade process. This includes more transparent communication of upcoming changes, a more gradual deprecation process, and a commitment to maintaining backward compatibility for a specified period.

Community-driven development
The development of Pandas 2.0 has been a collaborative effort, with input and contributions from the wider Pandas community. This has resulted in a more diverse set of features and improvements, addressing the needs of a broader range of users and use cases.

Pandas 2.0 represents a significant evolution of the library, building on its strengths while addressing long-standing limitations and challenges. The result is a more powerful, flexible, and user-friendly tool for data manipulation and analysis in Python. With its improved performance, enhanced type system, new APIs, better interoperability, and modernized codebase, Pandas 2.0 is well-suited to tackle the increasingly complex data analysis requirements of today's data-driven world.

Pandas Data Structures

Pandas provides two primary data structures for data manipulation and analysis: Series and DataFrame. Both of these data structures are built on top of NumPy arrays, which enable efficient and high-performance operations.

The given below is a detailed explanation of Series and DataFrame and how they work.

Series

A Series is a one-dimensional, labeled array capable of holding any data type, including integers, floats, strings, and Python objects. The basic structure of a Series consists of two components: data and index.

Data
The data in a Series is a one-dimensional NumPy array that holds the actual data values. The data can be of various types, such as integers, floats, or strings.

Index
The index is an array-like structure that labels each element in the data array. By default, the index is a zero-based integer sequence, but you can also provide custom labels for the index. This makes it easy to reference and manipulate data elements based on their labels.

To create a Series, you can use the pd.Series() constructor and pass in a list, dictionary, or NumPy array as data. You can also provide an optional index parameter to specify custom labels for the data elements.

DataFrame

A DataFrame is a two-dimensional, size-mutable, and heterogeneous tabular data structure with labeled axes (rows and columns). It can be thought of as a collection of Series objects, where each column is a Series. DataFrames are designed to handle a wide variety of data types, making them versatile and powerful tools for data manipulation and analysis.

The basic structure of a DataFrame consists of three components: data, index, and columns.

Data

The data in a DataFrame is stored as a collection of one or more Series objects, where each column represents a separate Series. The data in each Series can be of various types, such as integers, floats, or strings.

Index

The index is an array-like structure that labels each row in the DataFrame. Like the Series index, the DataFrame index can be a zero-based integer sequence or have custom labels.

Columns

The columns are an array-like structure that labels each column in the DataFrame. Column labels are used to reference and manipulate specific columns in the DataFrame.

To create a DataFrame, you can use the pd.DataFrame() constructor and pass in a variety of data types, such as a dictionary, a list of dictionaries, a 2D NumPy array, or even a list of Series. You can also provide optional index and columns parameters to specify custom row and column labels.

Both Series and DataFrame provide a wide range of methods and functionality for data manipulation and analysis. These include:
- Data selection: You can select specific rows, columns, or elements in a DataFrame or Series using indexing, slicing, and boolean conditions.
- Data cleaning: You can handle missing data, drop unwanted rows or columns, and replace specific values in a DataFrame or Series.
- Data transformation: You can apply functions to elements in a DataFrame or Series, create new columns, or reshape the data structure.
- Data aggregation and summarization: You can perform various aggregation functions, like sum, mean, or count, on a DataFrame or Series. You can also use

the GroupBy functionality to group data based on specific criteria and perform aggregation operations on each group.

- Sorting and ranking: You can sort a DataFrame or Series based on index or column values, and rank data elements based on specific criteria.
- Time series functionality: You can work with time series data, resample data at different frequencies, and perform rolling window calculations on a DataFrame or Series.

By understanding the structure and functionality of Series and DataFrame, you can effectively manipulate and analyze data in Pandas, allowing you to tackle a wide range of data-related tasks.

Load and Modify Dataset

To access and modify the data from the provided CSV file, follow these step-by-step instructions:

Download CSV File

First, download the Customer_Churn_Modelling.csv file from the GitHub repository to your local machine. You can do this by visiting the following URL in your web browser: https://raw.githubusercontent.com/kittenpub/database-repository/main/Customer_Churn_Modelling.csv

Right-click on the page and select "Save As" to save the file to your desired location.

Load Data

To load the data from the CSV file, you'll need to use the pd.read_csv() function in Pandas:

```python
import pandas as pd

# Replace 'path/to/your/csv' with the actual file path on your local machine
file_path = 'path/to/your/csv/Customer_Churn_Modelling.csv'
data = pd.read_csv(file_path)
```

Inspect the Data

Before modifying the data, it's essential to understand its structure and contents. You can use various Pandas functions to inspect the DataFrame:

```
# Display the first few rows of the DataFrame
print(data.head())

# Display the shape (number of rows and columns) of the DataFrame
print(data.shape)

# Display the data types of each column in the DataFrame
print(data.dtypes)

# Display summary statistics for numerical columns
print(data.describe())
```

Access and Modify Data

Now that you have an understanding of the data, you can access and modify it using various Pandas functions and methods. The following are some examples:

Select specific columns

```
# Select the 'Surname' column
surname = data['Surname']

# Select multiple columns, like 'CreditScore' and 'Age'
selected_columns = data[['CreditScore', 'Age']]
```

Select specific rows using slicing

```
# Select the first 10 rows of the DataFrame
first_10_rows = data[:10]
```

Select rows based on conditions

```python
# Select rows where 'Exited' equals 1 (churned customers)
churned_customers = data[data['Exited'] == 1]
```

Modify data values

```python
# Update the 'CreditScore' column by adding 10 points to each value
data['CreditScore'] = data['CreditScore'] + 10
```

Add a new column

```python
# Add a new column 'CreditScore_Categories' based on the 'CreditScore' values
def credit_score_category(score):
    if score < 300:
        return 'Bad'
    elif score < 600:
        return 'Poor'
    elif score < 700:
        return 'Fair'
    elif score < 800:
        return 'Good'
    else:
        return 'Excellent'

data['CreditScore_Categories'] =
data['CreditScore'].apply(credit_score_category)
```

Drop a column:

```python
# Drop the 'RowNumber' column
data = data.drop('RowNumber', axis=1)
```

Save the modified DataFrame to a new CSV file:

```python
# Save the modified DataFrame to a new CSV file
data.to_csv('Modified_Customer_Churn_Modelling.csv', index=False)
```

These are just a few examples of how to access and modify the data using Pandas for complex data manipulation tasks.

Summary

In this chapter, we discussed the powerful Python library Pandas and its role in data manipulation and analysis. We began with a brief introduction to Pandas and its significance in handling complex datasets efficiently. We then went through the process of installing and configuring Pandas, emphasizing the need to have Python and pip already installed.

We explored the version history of Pandas, starting from its initial release (0.1.x) to the latest major release (2.0). We highlighted the key improvements in Pandas 2.0, such as performance enhancements, a refined type system, new APIs, better interoperability, a simplified API, a modernized codebase, improved documentation, enhanced error messages, a structured deprecation policy, and a community-driven development process.

Next, we delved into the primary data structures in Pandas: Series and DataFrame. We provided a detailed explanation of their structure, components, and how they work. We emphasized that both Series and DataFrame offer a wide range of methods and functionality for data manipulation and analysis, making them essential tools for data science tasks.

Lastly, we went through a step-by-step walkthrough on accessing and modifying data from a CSV file using Pandas. We demonstrated how to download the file, load it using Pandas, inspect its structure and contents, and access and modify the data using various functions and methods. We also showed how to save the modified DataFrame to a new CSV file.

CHAPTER 2: DATA READ, STORAGE, AND FILE FORMATS

Reading CSV, Excel and JSON Data

Pandas provides several functions to read data from multiple formats, making it a versatile tool for working with different data sources. Some of the popular data formats supported by Pandas are CSV, JSON, Excel (XLS and XLSX), SQL, HTML, and more.

Now, let us assume the Customer_Churn_Modelling dataset is available in JSON and XLS formats. We will go through step-by-step instructions to read the data from these formats.

Download Files in JSON and XLS Formats

First, you'll need to have the Customer_Churn_Modelling dataset in JSON and XLS formats. You can convert the existing CSV file using an online converter or use a tool like Microsoft Excel or LibreOffice Calc to save the data in the desired format.

Read Data using Pandas

Reading JSON data

To read data from a JSON file, you can use the pd.read_json() function in Pandas:

```
import pandas as pd

# Replace 'path/to/your/json' with the actual file path on your local machine
json_file_path = 'path/to/your/json/Customer_Churn_Modelling.json'
data_json = pd.read_json(json_file_path)
```

Reading Excel (XLS or XLSX) data

To read data from an Excel file, you can use the pd.read_excel() function in Pandas. Note that you'll need to have the openpyxl package installed for reading XLSX files and xlrd package for reading XLS files.

```
# Install openpyxl and xlrd using pip
pip install openpyxl xlrd
import pandas as pd

# Replace 'path/to/your/xls' with the actual file path on your local machine
```

```
xls_file_path = 'path/to/your/xls/Customer_Churn_Modelling.xls'
data_xls = pd.read_excel(xls_file_path)
```

Inspect the Data

After loading the data, you can inspect the DataFrames to ensure that the data has been read correctly:

```
# Display the first few rows of the JSON DataFrame
print(data_json.head())

# Display the first few rows of the Excel DataFrame
print(data_xls.head())
```

Perform Data Manipulation and Analysis

Once the data is loaded, you can use the same Pandas functions and methods discussed earlier to manipulate and analyze the data, regardless of the original file format.

In summary, Pandas provides a seamless way to read data from various file formats, enabling users to work with different data sources without any hassle. By using appropriate functions like pd.read_json() and pd.read_excel(), you can load data from JSON and Excel files into DataFrames and perform data manipulation and analysis tasks as needed.

Writing Data to Different Formats

Pandas allows you to write DataFrames to different file formats easily. In this example, we'll assume you have already loaded the Customer_Churn_Modelling.csv dataset into a DataFrame called data. We'll now demonstrate how to write the data to various formats such as JSON, Excel (XLSX), and HTML.

Write to JSON Format

To write the DataFrame to a JSON file, you can use the to_json() method:

```
# Replace 'path/to/your/output' with the actual output file path on your local machine
json_output_path =
```

```
'path/to/your/output/Customer_Churn_Modelling_output.json'
data.to_json(json_output_path)
```

Write to Excel (XLSX) Format

To write the DataFrame to an Excel file, you can use the to_excel() method. Make sure you have the openpyxl package installed to write XLSX files.

```
# Install openpyxl using pip
pip install openpyxl
# Replace 'path/to/your/output' with the actual output file path on your local machine
xlsx_output_path =
'path/to/your/output/Customer_Churn_Modelling_output.xlsx'
data.to_excel(xlsx_output_path, index=False, engine='openpyxl')
```

Write to HTML Format

To write the DataFrame to an HTML file, you can use the to_html() method:

```
# Replace 'path/to/your/output' with the actual output file path on your local machine
html_output_path =
'path/to/your/output/Customer_Churn_Modelling_output.html'
data.to_html(html_output_path, index=False)
```

These are just a few examples of the output formats supported by Pandas. You can also write DataFrames to other formats like Parquet, Stata, and more. To write to these formats, you'll need to use the respective methods like to_parquet(), to_stata(), etc., and have the necessary packages installed.

In summary, Pandas provides an easy way to write DataFrames to various file formats, making it simple to share and store your data. By using the appropriate methods like to_json(), to_excel(), and to_html(), you can save the data from a DataFrame to different formats as required.

Database Interaction with Pandas 2.0

Database interaction refers to the process of connecting, querying, and managing data stored in a database management system (DBMS) using a programming language or software tool. Pandas, in combination with additional libraries, can interact with various databases, such as SQL databases (e.g., SQLite, MySQL, PostgreSQL) and NoSQL databases (e.g., MongoDB), to read and write data.

In this example, we'll demonstrate how to interact with an SQLite database using Pandas 2.0. SQLite is a lightweight, serverless, self-contained SQL database engine that's easy to set up and requires no separate server process.

Install sqlite3 Package

First, you'll need to install the sqlite3 package, which is part of Python's standard library and should already be installed.

```
!pip install pysqlite3
```

This code will install the sqlite3 package and its dependencies using pip

Setup SQLite Database

For demonstration purposes, we'll create a new SQLite database and a table to store the Customer_Churn_Modelling data.

```python
import sqlite3

# Connect to the SQLite database (this will create a new file 'churn.db' if it
doesn't exist)
conn = sqlite3.connect('churn.db')

# Create a cursor object to execute SQL commands
cursor = conn.cursor()

# Create the 'customer_churn' table
create_table_query = '''
```

```python
CREATE TABLE IF NOT EXISTS customer_churn (
    RowNumber INTEGER PRIMARY KEY,
    CustomerId INTEGER,
    Surname TEXT,
    CreditScore INTEGER,
    Geography TEXT,
    Gender TEXT,
    Age INTEGER,
    Tenure INTEGER,
    Balance REAL,
    NumOfProducts INTEGER,
    HasCrCard INTEGER,
    IsActiveMember INTEGER,
    EstimatedSalary REAL,
    Exited INTEGER
);
'''

cursor.execute(create_table_query)


# Commit the changes and close the cursor
conn.commit()
cursor.close()
```

Load Data into SQLite Database

Now, we'll insert the Customer_Churn_Modelling data (assumed to be in a DataFrame called data) into the 'customer_churn' table.

```python
# Insert the data from the DataFrame into the SQLite table
data.to_sql('customer_churn', conn, if_exists='replace', index=False)
```

Query SQLite database using Pandas

You can use Pandas to execute SQL queries on the database and store the results in a DataFrame.

```python
# Define the SQL query
sql_query = 'SELECT * FROM customer_churn WHERE Exited = 1'

# Execute the query and store the results in a DataFrame
churned_customers = pd.read_sql_query(sql_query, conn)
```

Perform Data Manipulation and Analysis

Once the data is loaded into a DataFrame, you can use the same Pandas functions and methods discussed earlier to manipulate and analyze the data.

Write Data Back to SQLite Database (optional)

If you want to write modified data back to the SQLite database, you can use the to_sql() method again.

```python
# Assuming 'modified_data' is a DataFrame containing modified data
modified_data.to_sql('customer_churn', conn, if_exists='replace', index=False)
```

Close the Connection

Finally, close the connection to the SQLite database.

```python
conn.close()
```

In summary, Pandas allows you to interact with databases by connecting to them, querying data, and managing the data within the database. By using additional libraries, like sqlite3 in this example, you can perform database interaction tasks and use Pandas to read and write data from and to the database. This integration makes Pandas a powerful tool for data manipulation and analysis involving databases.

Connect Web APIs

Utilizing web APIs through Pandas enables the acquisition of up-to-the-minute information, which can then be seamlessly integrated into a DataFrame for subsequent evaluation and manipulation. This integration with APIs streamlines the process of accessing and working with data in real-time, providing a more efficient approach to analysis. In this example, we'll demonstrate how to connect to a public API using the requests library and load the data into a Pandas DataFrame.

Install HTTP Requests Package

First, you'll need to install the requests package for making HTTP requests:

```
pip install requests
```

Make API Request and Retrieve Data

For this example, we'll use the public API provided by JSONPlaceholder, which returns a list of dummy posts in JSON format.

```python
import requests

# Define the API endpoint
api_url = 'https://jsonplaceholder.typicode.com/posts'

# Make an HTTP GET request to the API
response = requests.get(api_url)

# Check if the request was successful
if response.status_code == 200:
    # Parse the JSON data
    json_data = response.json()
else:
    print(f'Error: Unable to fetch data from the API (status code: {response.status_code})')
```

Load JSON Data into Pandas DataFrame

Once you have the JSON data, you can load it into a DataFrame using the pd.DataFrame() constructor or the pd.read_json() function.

```python
import pandas as pd

# Load the JSON data into a DataFrame
data = pd.DataFrame(json_data)

# Alternatively, you can use the 'pd.read_json()' function
# data = pd.read_json(api_url)
```

Perform Data Manipulation and Analysis

Now that the data is loaded into a DataFrame, you can use the same Pandas functions and methods discussed earlier to manipulate and analyze the data.

Keep in mind that the process for connecting to different APIs may vary depending on the specific API and the required authentication methods. You may need to provide API keys, tokens, or use OAuth for authentication. Be sure to consult the API documentation for detailed instructions on how to connect and retrieve data.

In summary, you can connect to web APIs and retrieve real-time data using libraries like requests and load the data into Pandas DataFrames for manipulation and analysis. This capability makes Pandas a powerful tool for working with dynamic data sources and real-time applications.

Data Scraping

Extracting structured data from websites is achieved through a process known as data scraping. This involves parsing the HTML content of a website to gather information. Beautiful Soup, a Python library, is a commonly used tool for web scraping. With its simple and user-friendly syntax, it allows users to quickly and easily navigate and parse HTML and XML documents. Beautiful Soup can be used to extract various types of data from websites, such as text, links, and images. It also offers support for multiple parsers, making it compatible with different markup languages. In this example, we'll demonstrate how to use Beautiful Soup to scrape data from a website and load it into a Pandas DataFrame.

<u>Install Beautful Soup Package</u>

First, you'll need to install Beautiful Soup and the requests library:

```
pip install beautifulsoup4 requests
```

Fetch HTML Content from Website

For this example, we'll scrape data from a simple HTML table available at the following URL:
https://www.worldcoinindex.com/

```python
import requests

# Define the URL
url = 'https://www.worldcoinindex.com/'

# Fetch the HTML content
response = requests.get(url)

# Check if the request was successful
if response.status_code == 200:
    html_content = response.text
else:
    print(f'Error: Unable to fetch data from the website (status code: {response.status_code})')
```

Parse HTML Content using Beautiful Soup

Next, we'll parse the HTML content using Beautiful Soup to extract the data we need.

```python
from bs4 import BeautifulSoup

# Parse the HTML content
soup = BeautifulSoup(html_content, 'html.parser')
```

```python
# Find the table containing the data we want to scrape
table = soup.find('table', {'class': 'table'})
```

Extract Data from HTML Table

Now, we'll extract the data from the HTML table, including the headers and row data.

```python
# Extract the headers
headers = [header.text for header in table.findAll('th')]

# Extract the rows
rows = []
for row in table.tbody.findAll('tr'):
    rows.append([col.text for col in row.findAll('td')])
```

Load Data into Pandas DataFrame

Once we have the headers and rows, we can load the data into a DataFrame.

```python
import pandas as pd

# Create a DataFrame using the extracted headers and rows
data = pd.DataFrame(rows, columns=headers)
```

Perform Data Manipulation and Analysis

Now that the data is loaded into a DataFrame, you can use the same Pandas functions and methods discussed earlier to manipulate and analyze the data.

In summary, you can perform web scraping using libraries like Beautiful Soup and Requests to extract structured data from websites. Once you have the data, you can load it into a Pandas DataFrame for further processing and analysis. This capability makes Pandas a powerful tool for working with data from various sources, including dynamic web content.

Handling Missing Data

In the data cleaning process, it is crucial to address any missing data. Pandas offers a range of techniques to identify, evaluate, and manage missing data. These methods can help to ensure that the data analysis is accurate and reliable by filling in the missing values or removing the incomplete records. In this example, we'll assume you have already loaded the Customer_Churn_Modelling.csv dataset into a DataFrame called data.

Identify Missing Data

First, use the isna() or isnull() method to detect missing values in the DataFrame. The isna() method returns a DataFrame of the same shape with True for missing values and False for non-missing values.

```
missing_values = data.isna()
```

You can use the sum() method to count the number of missing values per column:

```
missing_value_counts = data.isna().sum()
print(missing_value_counts)
```

Handle Missing Data

There are various ways to handle missing data in a DataFrame, such as dropping or filling missing values. The choice depends on the nature of the data and the specific use case.

Drop missing values

The dropna() method removes rows or columns with missing values from the DataFrame. By default, it removes any row containing at least one missing value:

```
data_cleaned = data.dropna()
```

To drop columns with missing values instead of rows, set the axis parameter to 1:

```
data_cleaned = data.dropna(axis=1)
```

Fill missing values

The fillna() method fills missing values with a specified value or using a specified method (such as forward fill or backward fill).

To fill missing values with a constant value (e.g., 0):

```
data_filled = data.fillna(0)
```

To fill missing values with the mean of the respective column:

```
data_filled = data.fillna(data.mean())
```

To fill missing values using forward fill (propagate the last valid observation forward):

```
data_filled = data.fillna(method='ffill')
```

To fill missing values using backward fill (propagate the next valid observation backward):

```
data_filled = data.fillna(method='bfill')
```

Verify the Changes

After handling the missing data, you can use the isna().sum() method again to confirm that there are no more missing values in the DataFrame:

```
missing_value_counts_cleaned = data_cleaned.isna().sum()
print(missing_value_counts_cleaned)
```

In summary, Pandas provides several methods to handle missing data in DataFrames. By using methods like dropna() and fillna(), you can efficiently handle missing values based on your specific requirements and the nature of the data.

Data Transformation and Cleaning

Effective data preprocessing involves crucial stages of data transformation and cleaning. These processes ensure that the collected data is appropriately organized, consistent, and reliable, thereby leading to better analysis and decision-making. It is essential to conduct these initial steps as they significantly impact the accuracy and quality of the final output. Incomplete or erroneous data can lead to incorrect conclusions and negatively impact the results.

In this example, we'll use the Customer_Churn_Modelling.csv dataset, assuming it's already loaded into a DataFrame called data.

Identify and Handle Missing Data

Refer to the previous section on how to handle missing data with Pandas. Apply either the dropna() or fillna() method based on your specific requirements.

Rename Columns (optional)

If you want to rename columns for better readability or consistency, you can use the rename() method:

```python
data = data.rename(columns={
    'NumOfProducts': 'Number_of_Products',
    'HasCrCard': 'Has_Credit_Card',
    'IsActiveMember': 'Is_Active_Member',
    'EstimatedSalary': 'Estimated_Salary'
})
```

Convert Data Types

Ensure that the data types of each column are appropriate for further analysis. You can check the data types using the dtypes attribute:

```python
print(data.dtypes)
```

If you need to change the data type of a column, you can use the astype() method:

```python
data['Has_Credit_Card'] = data['Has_Credit_Card'].astype('bool')
data['Is_Active_Member'] = data['Is_Active_Member'].astype('bool')
```

Remove Duplicates

Check for and remove any duplicate rows in the dataset using the duplicated() and drop_duplicates() methods:

```python
duplicates = data.duplicated()
print(f'Number of duplicate rows: {duplicates.sum()}')


data = data.drop_duplicates()
```

Normalize Numerical Columns

Normalization and standardization help scale numerical features to a common range or distribution, which can be useful for certain machine learning algorithms.

To normalize a column (scale values between 0 and 1):

```python
data['CreditScore'] = (data['CreditScore'] - data['CreditScore'].min()) / (data['CreditScore'].max() - data['CreditScore'].min())
```

To standardize a column (scale values to have a mean of 0 and a standard deviation of 1):

```python
data['Age'] = (data['Age'] - data['Age'].mean()) / data['Age'].std()
```

Encode Categorical Variables

If you plan to use the data for machine learning, you may need to encode categorical variables as numerical values. You can use techniques like one-hot encoding or label encoding.

For one-hot encoding, use the get_dummies() function:

```python
data = pd.get_dummies(data, columns=['Geography', 'Gender'])
```

For label encoding, you can use the LabelEncoder class from scikit-learn:

```python
from sklearn.preprocessing import LabelEncoder
```

```python
encoder = LabelEncoder()
data['Surname'] = encoder.fit_transform(data['Surname'])
```

Reorder or Drop Columns

Reorder or drop columns as needed for better organization or to remove unnecessary features:

```python
# Reorder columns
data = data[['RowNumber', 'CustomerId', 'Surname', 'CreditScore', 'Age',
'Tenure', 'Balance', 'Number_of_Products', 'Has_Credit_Card',
'Is_Active_Member', 'Estimated_Salary', 'Geography_France',
'Geography_Germany', 'Geography_Spain', 'Gender_Female', 'Gender_Male',
'Exited']]

# Drop unnecessary columns (e.g
Drop unnecessary columns (e.g., 'RowNumber', 'CustomerId', 'Surname' might
not be relevant for your analysis)
data = data.drop(columns=['RowNumber', 'CustomerId', 'Surname'])
```

Verify the Changes

After performing data transformation and cleaning, it's essential to verify the changes by inspecting the data:

```python
print(data.head())
print(data.dtypes)
print(data.describe())
```

In summary, you can use various methods provided by Pandas to perform data transformation and cleaning tasks, such as handling missing values, renaming columns, converting data types, removing duplicates, normalizing or standardizing numerical columns, encoding categorical variables, and reordering or dropping columns. These steps help ensure the data is accurate, consistent, and well-structured for further analysis or machine learning applications.

Summary

We discussed reading data from different file formats using Pandas. Pandas can read multiple formats like CSV, JSON, Excel, SQL databases, and more. We demonstrated how to read data from the Customer_Churn_Modelling.csv dataset in JSON and Excel (XLS) formats using the pd.read_json() and pd.read_excel() functions.

Next, we covered writing data to different formats. Pandas can write DataFrames to various file formats, including CSV, JSON, Excel, and more. We demonstrated how to write the DataFrame containing the Customer_Churn_Modelling.csv dataset to a JSON and Excel file using the to_json() and to_excel() methods.

Following that, we discussed database interaction with Pandas. Pandas allows you to perform database operations using the pd.read_sql() and to_sql() functions. We demonstrated how to connect to a SQLite database, read data from a table into a DataFrame using pd.read_sql(), and write data from a DataFrame to a table using the to_sql() method.

We then explored connecting web APIs to Pandas. By using the requests library to fetch real-time data from APIs and loading it into a DataFrame, you can manipulate and analyze dynamic data sources. We provided an example of fetching data from the JSONPlaceholder API and loading it into a DataFrame using the pd.DataFrame() constructor.

Afterward, we discussed web scraping using Beautiful Soup. By fetching HTML content from a website, parsing it with Beautiful Soup, extracting structured data, and loading it into a DataFrame, you can work with data from dynamic web content. We provided an example of scraping data from a table on the World Coin Index website and loading it into a DataFrame.

Next, we explained how to handle missing data in Pandas. The isna() and isnull() methods help identify missing values, while the dropna() and fillna() methods help handle them by either removing or filling missing values based on your requirements.

Finally, we covered data transformation and cleaning using Pandas. This process involves handling missing data, renaming columns, converting data types, removing duplicates, normalizing or standardizing numerical columns, encoding categorical variables, and reordering or dropping columns. These steps ensure that the data is accurate, consistent, and well-structured for further analysis or machine learning applications.

Overall, this chapter has covered various aspects of using Pandas for data manipulation, including reading and writing data from different formats, database interaction, connecting to web APIs and web scraping, handling missing data, and performing data transformation and cleaning.

Chapter 3: Indexing and Selecting Data

Basics of Indexing and Selection

Efficient data analysis and manipulation rely heavily on the ability to access, filter, and modify specific data points, rows, or columns within a DataFrame. Pandas provides an array of essential tools for performing these fundamental operations, such as indexing and selection. These tools not only allow for the quick and easy retrieval of specific data, but they also enable data cleaning and transformation tasks to be performed with greater accuracy and speed. With Pandas' robust indexing and selection capabilities, data analysts and scientists can efficiently handle complex data sets and extract insights that drive informed decision-making.

In this example, we'll assume you have already loaded the Customer_Churn_Modelling.csv dataset into a DataFrame called data.

Selecting Columns

You can select a single column using the column name:

```
credit_scores = data['CreditScore']
```

Or multiple columns by providing a list of column names:

```
selected_columns = data[['CreditScore', 'Age', 'Exited']]
```

Selecting Rows by Index

You can select rows using the iloc[] and loc[] properties. The iloc[] property is used for integer-based indexing, while the loc[] property is used for label-based indexing.

Using iloc[] to select the first row:

```
first_row = data.iloc[0]
```

Using iloc[] to select multiple rows:

```
rows_2_to_4 = data.iloc[1:4]
```

Using loc[] to select a specific row by index label (assuming the DataFrame has a custom index):

```python
row_with_label = data.loc['index_label']
```

Selecting Specific Data Points

You can combine row and column selections to access specific data points.

Using iloc[] to select the value in the first row and first column:

```python
first_value = data.iloc[0, 0]
```

Using loc[] to select a value by row and column labels (assuming the DataFrame has a custom index):

```python
specific_value = data.loc['index_label', 'CreditScore']
```

Conditional Selection

You can use boolean conditions to filter rows based on certain criteria.

For example, to select rows where 'Exited' is equal to 1:

```python
exited_rows = data[data['Exited'] == 1]
```

You can also combine multiple conditions using the & (and) or | (or) operators:

```python
exited_and_has_credit_card = data[(data['Exited'] == 1) &
(data['Has_Credit_Card'] == True)]
```

Setting Custom Index

You can set a custom index using the set_index() method. For example, to set the 'CustomerId' column as the index:

```python
data = data.set_index('CustomerId')
```

Remember that this operation creates a new DataFrame with the custom index. If you want to modify the existing DataFrame, set the inplace parameter to True:

```
data.set_index('CustomerId', inplace=True)
```

In summary, indexing and selection are fundamental operations in Pandas that allow you to access and manipulate specific data points, rows, or columns within a DataFrame. By using methods like iloc[], loc[], and conditional selections, you can perform various data analysis, cleaning, and transformation tasks efficiently.

Multi-Indexing and Hierarchical Indexing

Hierarchical indexing, also referred to as multi-indexing, is a feature of Pandas that facilitates the creation of multiple levels of index in a single DataFrame. This functionality is particularly useful when dealing with complex, multi-dimensional datasets, as it allows for more organized and natural handling of the data. By setting up a hierarchical index, users can easily slice and manipulate subsets of data based on specific index levels, allowing for more precise and efficient analysis. In addition, hierarchical indexing can help to reduce the need for reshaping and restructuring of data, which can save time and effort in the data preparation phase of analysis.

Let us assume you've already loaded the Customer_Churn_Modelling.csv dataset into a DataFrame called data. We'll create a multi-level index based on the 'Geography' and 'Gender' columns.

Creating Multi-level Index

You can create a multi-level index using the set_index() method and providing a list of columns:

```
data_multi = data.set_index(['Geography', 'Gender'])
```

Now, the DataFrame has a multi-level index based on 'Geography' and 'Gender'. To view the first few rows of the DataFrame:

```
print(data_multi.head())
```

Accessing Data with Multi-level Index

You can use the loc[] property to access data within the multi-level index. For example, to access all data for the 'France' and 'Female' group:

```
france_female = data_multi.loc[('France', 'Female')]
```

Slicing with Multi-level Index

Using the slice(None) function, you can perform slicing operations on a multi-level index. For example, to select all rows where 'Geography' is 'Germany':

```
germany_data = data_multi.loc[(slice('Germany', 'Germany'), slice(None)), :]
```

Selecting Data at Specific Level

You can use the xs() method to select data at a specific level. For example, to select all data where 'Gender' is 'Male':

```
male_data = data_multi.xs('Male', level='Gender', axis=0)
```

Swapping Levels

You can swap levels of a multi-level index using the swaplevel() method. For example, to swap the 'Geography' and 'Gender' levels:

```
data_swapped = data_multi.swaplevel('Geography', 'Gender')
```

Sorting Multi-level Index

You can sort the DataFrame based on index levels using the sort_index() method. To sort the DataFrame by the 'Geography' and 'Gender' levels:

```
data_sorted = data_multi.sort_index(level=['Geography', 'Gender'])
```

Resetting Multi-level Index

To reset a multi-level index and convert it back to columns, use the reset_index() method:

```python
data_flat = data_multi.reset_index()
```

In summary, multi-indexing or hierarchical indexing in Pandas allows you to have multiple index levels within a single DataFrame, providing a more structured and intuitive way to handle higher-dimensional data. By creating a multi-level index with the set_index() method and using various methods like loc[], xs(), swaplevel(), and sort_index(), you can efficiently manipulate and access data within a multi-level index.

Indexing with Booleans and Conditional Selection

Indexing with Booleans and Conditional Selection, enables filtering of rows in a DataFrame based on specific conditions. These techniques are valuable for a range of data analysis, cleaning, and transformation tasks. By using Boolean indexing, you can select rows of data that satisfy specific criteria, such as selecting all rows where a particular column meets a certain condition. Similarly, conditional selection allows you to filter data based on more complex criteria, such as selecting all rows where multiple columns meet specific conditions. These techniques are powerful tools for data manipulation and can be applied in various contexts, such as data preprocessing, exploratory data analysis, and machine learning.

We'll assume you have already loaded the Customer_Churn_Modelling.csv dataset into a DataFrame called data.

Boolean Indexing

Boolean indexing involves creating a boolean mask, a Pandas Series with the same index as the DataFrame and containing True or False values based on a condition. When this mask is applied to the DataFrame, only the rows with True values are selected.

For example, let us create a boolean mask for rows where the 'Exited' column is equal to 1:

```python
mask = data['Exited'] == 1
```

Now, apply the boolean mask to the DataFrame to filter the rows:

```python
exited_customers = data[mask]
```

You can also apply the mask directly:

```python
exited_customers = data[data['Exited'] == 1]
```

Conditional Selection with Multiple Conditions

You can use the & (and) or | (or) operators to combine multiple conditions.

For example, to select rows where 'Exited' is equal to 1 and 'Age' is greater than 40:

```python
exited_and_age_above_40 = data[(data['Exited'] == 1) & (data['Age'] > 40)]
```

Or, to select rows where 'Exited' is equal to 1 or 'Age' is greater than 40:

```python
exited_or_age_above_40 = data[(data['Exited'] == 1) | (data['Age'] > 40)]
```

Using query() Method for Conditional Selection

Another way to perform conditional selection is by using the query() method. This method allows you to write conditions as a string expression. For example, to select rows where 'Exited' is equal to 1 and 'Age' is greater than 40:

```python
exited_and_age_above_40 = data.query("Exited == 1 and Age > 40")
```

In summary, Boolean indexing and conditional selection in Pandas allow you to filter rows in a DataFrame based on specific conditions. By creating boolean masks, combining multiple conditions with & or | operators, or using the query() method, you can perform various data analysis, cleaning, and transformation tasks more efficiently.

Using .loc, .iloc, and .at

The .loc[], .iloc[], and .at[] properties in Pandas are used for accessing and selecting data in DataFrames based on different criteria. They are useful for data analysis, cleaning, and transformation tasks.

Let us assume you have already loaded the Customer_Churn_Modelling.csv dataset into a DataFrame called data.

.loc[] Property

The .loc[] property is used for label-based indexing. You can use it to select data by row and column labels. For example, assuming you have set the 'CustomerId' column as the index:

```
data = data.set_index('CustomerId')
```

To select data for a specific credit score:

```
customer_data = data.loc[15745662, 'CreditScore']
```

To select multiple rows and columns using the .loc[] property, you can pass in lists or slices of labels:

```
selected_customers = data.loc[[15745662, 15734523], ['CreditScore', 'Age']]
selected_rows_columns = data.loc[15745662:15734523, 'CreditScore':'Age']
```

.iloc[] Property

The .iloc[] property is used for integer-based (positional) indexing. You can use it to select data by row and column positions.

For example, to select data from the first row and the second column:

```
first_row_second_column = data.iloc[0, 1]
```

To select multiple rows and columns using the .iloc[] property, you can pass in lists or slices of positions:

```
selected_rows = data.iloc[[0, 1, 2], :]
selected_columns = data.iloc[:, [0, 1, 2]]
selected_rows_columns = data.iloc[0:3, 0:3]
```

.at[] Property

The .at[] property is used for fast label-based indexing of scalar values. It is similar to .loc[]

but is optimized for accessing single data points, making it faster for this specific use case.

For example, to select data for a specific customer ID and column:

```
customer_data = data.at[15745662, 'CreditScore']
```

Keep in mind that .at[] can only be used for accessing single data points and not for selecting multiple rows or columns.

In summary, the .loc[], .iloc[], and .at[] properties in Pandas are used for accessing and selecting data in DataFrames based on different criteria: label-based indexing for .loc[] and .at[], and integer-based (positional) indexing for .iloc[]. By using these properties, you can efficiently perform various data analysis, cleaning, and transformation tasks.

Slicing and Subsetting Data

Pandas offers slicing and subsetting techniques that enable users to extract particular rows, columns, or data points within a DataFrame based on their positions or index labels. These methods are valuable for tasks such as data analysis, cleaning, and transformation. By using slicing, one can slice a DataFrame and select the desired rows and columns. Subsetting is another technique that involves creating a new DataFrame based on a subset of the original data. With these methods, users can easily isolate relevant data and perform necessary modifications without affecting.

Now, let us assume you have already loaded the Customer_Churn_Modelling.csv dataset into a DataFrame called data.

Slicing Rows

You can slice rows by their positions using the iloc[] property with a start and end index.

For example, to select rows from index 5 to 9:

```
subset_rows = data.iloc[5:10]
```

Slicing Columns

You can also slice columns by their positions using the iloc[] property.

For example, to select the first three columns:

```
subset_columns = data.iloc[:, :3]
```

Slicing Rows and Columns

You can combine row and column slicing using the iloc[] property.

For example, to select rows from index 5 to 9 and the first three columns:

```
subset_data = data.iloc[5:10, :3]
```

Slicing with loc[] Property

If you want to slice based on index labels instead of positions, you can use the loc[] property.

For example, assuming you have set 'CustomerId' as the index, to select rows with index labels 15634602 to 15634606:

```
subset_data_labels = data.loc[15634602:15634606, :]
```

Subsetting Data based on Column Names

You can select specific columns by providing a list of column names.

For example, to select only the 'CreditScore', 'Age', and 'Exited' columns:

```
selected_columns = data[['CreditScore', 'Age', 'Exited']]
```

Subsetting Data with Cconditions (Boolean Indexing)

You can use Boolean indexing to select rows based on specific conditions.

For example, to select rows where 'Exited' is equal to 1:

```
exited_customers = data[data['Exited'] == 1]
```

You can use the query() method to filter rows based on specific conditions expressed as a string.

For example, to select rows where 'Exited' is equal to 1 and 'Age' is greater than 40:

```python
exited_and_age_above_40 = data.query("Exited == 1 and Age > 40")
```

In summary, slicing and subsetting data in Pandas allow you to select specific rows, columns, or data points within a DataFrame based on their positions, index labels, or conditions. By using methods like iloc[], loc[], Boolean indexing, or the query() method, you can perform various data analysis, cleaning, and transformation tasks more efficiently.

Modifying Data using Indexing and Selection

Pandas offers indexing and selection techniques that allow modifying the data within DataFrames beyond just selecting it. Besides selecting data, these techniques enable making changes to specific subsets of data within the DataFrame, such as updating or replacing values, adding or deleting columns or rows, and more. This flexibility in data manipulation helps in various data analysis tasks and allows users to perform complex operations efficiently. The following are some examples of how to modify data using indexing and selection methods:

Modify Single Value using .at[] Property

The .at[] property can be used to modify a single value in the DataFrame based on its row and column labels.

For example, assuming you have set the 'CustomerId' column as the index:
python

```python
data.at[15745662, 'CreditScore'] = 750
```

This will set the 'CreditScore' value for the customer with ID 15745662 to 750.

Modify Single Value using .iat[] Property

Similar to the .at[] property, you can use the .iat[] property to modify a single value in the DataFrame based on its row and column positions:

```python
data.iat[0, 0] = 750
```

This will set the value in the first row and first column to 750.

Modify Multiple Values in Column using Boolean Indexing

You can use boolean indexing to modify multiple values in a column based on a condition.

For example, to increase the 'CreditScore' of all customers older than 40 by 10 points:

```python
mask = data['Age'] > 40
data.loc[mask, 'CreditScore'] += 10
```

Modify Values in Column using .apply() Method

You can use the .apply() method to apply a custom function to each value in a column.

For example, to round the 'CreditScore' values to the nearest 10:

```python
import numpy as np

def round_to_nearest_ten(x):
    return np.round(x, -1)

data['CreditScore'] = data['CreditScore'].apply(round_to_nearest_ten)
```

Modify Values in Multiple Columns using .applymap() Method

You can use the .applymap() method to apply a custom function to each value in multiple columns.

For example, to round all numerical columns to the nearest integer:

```python
numerical_columns = ['CreditScore', 'Age', 'Tenure', 'Balance',
'NumOfProducts', 'EstimatedSalary']

data[numerical_columns] = data[numerical_columns].applymap(np.round)
```

In summary, you can use indexing and selection techniques in Pandas, such as .at[], .iat[], boolean indexing, and the .apply() and .applymap() methods, to modify data in DataFrames. These techniques are useful for various data analysis, cleaning, and transformation tasks.

Advanced Indexing Techniques

Cross-section (xs()) Method

The xs() method allows you to select data based on a specific level of a multi-indexed DataFrame. This can be useful when you have a DataFrame with hierarchical indexing. Let us create a multi-indexed DataFrame with 'Geography' and 'Gender' as index levels:

```python
multi_indexed_data = data.set_index(['Geography', 'Gender'])
```

Now, you can use the xs() method to select data for a specific level:

```python
german_customers = multi_indexed_data.xs('Germany', level='Geography')
```

where() Method

The where() method allows you to filter data based on conditions while keeping the original shape of the DataFrame. This can be useful when you want to preserve the DataFrame structure or visualize the data.

For example, to keep only rows where 'Exited' is equal to 1 and replace others with NaN:

```python
exited_customers = data.where(data['Exited'] == 1)
```

mask() Method

The mask() method works similarly to the where() method but inverts the condition. Instead of keeping the data points that meet the condition, it replaces them with NaN or a specified value.

For example, to replace the 'CreditScore' values of all customers who exited with NaN:

```
data['CreditScore'] = data['CreditScore'].mask(data['Exited'] == 1)
```

isin() Method

The isin() method allows you to filter data based on whether the values are in a given list or not. This can be useful when you want to select data for specific categories or groups.

For example, to select rows where 'Geography' is either 'France' or 'Germany':

```
selected_countries = data[data['Geography'].isin(['France', 'Germany'])]
```

eval() Method

The eval() method allows you to evaluate complex expressions involving columns of the DataFrame. This can be useful when you want to perform arithmetic operations or boolean conditions involving multiple columns.

For example, to calculate the ratio of 'Balance' to 'EstimatedSalary' for each customer:

```
data['BalanceToSalaryRatio'] = data.eval("Balance / EstimatedSalary")
```

In summary, advanced indexing techniques in Pandas such as xs(), where(), mask(), isin(), and eval() methods can help you access and manipulate data more efficiently in the given database. These techniques offer more flexibility and can be useful for various data analysis, cleaning, and transformation tasks.

Summary

We discussed different indexing and selection methods in Pandas, including .loc[], .iloc[], and .at[]. The .loc[] property is used for label-based indexing, allowing you to select data by row and column labels. The .iloc[] property is used for integer-based (positional) indexing, allowing you to select data by row and column positions. The .at[] property is used for fast label-based indexing of scalar values and is optimized for accessing single data points.

We then discussed modifying data using indexing and selection techniques. The .at[] and .iat[] properties can be used to modify single values in the DataFrame based on their row and column labels or positions, respectively. Boolean indexing allows you to modify multiple values in a column based on a condition. The .apply() method can be used to apply a custom function to each value in a column, while the .applymap() method can be used to apply a custom function to each value in multiple columns.

Next, we covered advanced indexing techniques in Pandas, including the xs(), where(), mask(), isin(), and eval() methods. The xs() method allows you to select data based on a specific level of a multi-indexed DataFrame. The where() method filters data based on conditions while keeping the original shape of the DataFrame. The mask() method works similarly to the where() method but inverts the condition, replacing data points that meet the condition with NaN or a specified value. The isin() method filters data based on whether the values are in a given list or not, useful for selecting data for specific categories or groups. The eval() method evaluates complex expressions involving columns of the DataFrame, useful for performing arithmetic operations or boolean conditions involving multiple columns.

These indexing, selection, and advanced techniques provide flexible and efficient ways to access, manipulate, and modify data in Pandas DataFrames. They are essential tools for data analysis, cleaning, and transformation tasks.

CHAPTER 4: DATA MANIPULATION AND TRANSFORMATION

Merging and Joining DataFrames

In Pandas, the process of combining two or more DataFrames based on common columns or indexes can be achieved through merging and joining. These operations play a critical role in managing multiple related datasets, allowing for efficient data analysis and manipulation. By merging DataFrames, data can be combined based on shared columns, resulting in a single DataFrame that includes all the relevant information. Joining DataFrames, on the other hand, enables the combination of two or more DataFrames based on their indexes. Both of these techniques can help simplify data management, improve data quality, and provide a better understanding of the relationships between datasets. Therefore, understanding how to merge and join DataFrames in Pandas is an important skill for data analysts and scientists.

For illustration purposes, let us assume you have another dataset containing customer income data. We'll create a small DataFrame called customer_income with the same 'CustomerId' column as the data DataFrame:

```python
import pandas as pd

customer_income = pd.DataFrame({
    'CustomerId': [15745662, 15734523, 15698723],
    'AnnualIncome': [55000, 68000, 45000]
})
```

Merging DataFrames using merge() Function

The merge() function combines DataFrames based on common columns. By default, it performs an inner join, which means that only rows with matching values in the common column(s) will be included in the result.

```python
merged_data = pd.merge(data, customer_income, on='CustomerId')
```

You can specify the type of join by using the how parameter:
- 'left': Use keys from the left DataFrame only
- 'right': Use keys from the right DataFrame only
- 'outer': Use keys from both DataFrames
- 'inner': Use intersection of keys from both DataFrames (default)

For example, to perform a left join:

```
merged_data_left = pd.merge(data, customer_income, on='CustomerId',
how='left')
```

Joining DataFrames using join() Method

The join() method is used to combine DataFrames based on their index. By default, it performs a left join, which means that it uses the index from the left DataFrame and includes all its rows.

First, let us set 'CustomerId' as the index for both DataFrames:

```
data = data.set_index('CustomerId')
customer_income = customer_income.set_index('CustomerId')
```

Now, you can join the DataFrames:

```
joined_data = data.join(customer_income)
```

You can also specify the type of join using the how parameter, just like with the merge() function.

In summary, merging and joining DataFrames in Pandas allow you to combine two or more DataFrames based on common columns or indexes. The merge() function combines DataFrames based on common columns, while the join() method combines DataFrames based on their index. These operations are essential when working with multiple related datasets and can be adjusted to perform various types of joins (inner, left, right, and outer) depending on your requirements.

Concatenating and Appending DataFrames

In Pandas, merging and adding DataFrames refer to the process of joining two or more DataFrames along a specific axis, either horizontally or vertically. This technique is beneficial when you have multiple datasets with the same rows or columns and want to stack them together. Concatenation and appending are two methods of combining data frames in Pandas. Concatenation is used to stack DataFrames on top of each other or next to each other, while appending adds new rows to an existing DataFrame. Both of these methods can be useful in creating a single DataFrame that contains all the necessary information from the original datasets.

For illustration purposes, let us assume you have two datasets, data1 and data2, which are subsets of the original data DataFrame:

```
data1 = data.head(10)
data2 = data.tail(10)
```

Concatenating DataFrames using concat() Function

The concat() function combines DataFrames along a particular axis. By default, it concatenates DataFrames along the rows (axis=0).

The given below teaches how to concatenate data1 and data2:

```
concatenated_data = pd.concat([data1, data2])
```

To concatenate DataFrames along the columns, set the axis parameter to 1:

```
concatenated_data_columns = pd.concat([data1, data2], axis=1)
```

Keep in mind that when concatenating DataFrames along columns, it's crucial that they have the same index. Otherwise, you might end up with missing values in the result.

Appending DataFrames using append() Method

The append() method is a convenient way to combine DataFrames along the rows, similar to concatenating along the rows with the concat() function. You can append data2 to data1 as follows:

```
appended_data = data1.append(data2)
```

Note that the append() method does not modify the original DataFrames; it creates a new DataFrame containing the combined data.

In summary, concatenating and appending DataFrames in Pandas allow you to combine two or more DataFrames along a particular axis (either rows or columns). The concat()

function combines DataFrames along a specified axis, while the append() method is a convenient way to combine DataFrames along the rows. These operations are useful when you have multiple datasets with the same columns or rows and want to stack them together.

Sample Program on Concatenate and Append

Let us create another example using the given Customer_Churn_Modelling.csv dataset. We'll first create two new DataFrames with different subsets of the original data and then concatenate and append them.

Split the original data into two subsets

```python
# Read the original data
data = pd.read_csv("https://raw.githubusercontent.com/kittenpub/database-repository/main/Customer_Churn_Modelling.csv")

# Create subsets of data for customers from France and Spain
data_france = data[data['Geography'] == 'France']
data_spain = data[data['Geography'] == 'Spain']
```

Concatenate the DataFrames

```python
# Concatenate the DataFrames along the rows (axis=0)
concatenated_data = pd.concat([data_france, data_spain])
```

Append the DataFrames

```python
# Append 'data_spain' to 'data_france'
appended_data = data_france.append(data_spain)
```

Both concatenated_data and appended_data will contain rows from the original DataFrame, where customers are from France and Spain. Although the results of concatenating and appending DataFrames along the rows are the same in this example, these techniques can be used in different scenarios depending on your requirements. Remember that concat() can be used to combine DataFrames along either rows or columns, while append() is specifically for combining DataFrames along the rows.

Pivoting and Melting DataFrames

Pivoting and melting DataFrames in Pandas involve reshaping your data by changing its structure without altering the actual data. Pivoting transforms the data from a long format to a wide format, while melting does the opposite.

Pivoting DataFrames using pivot() Method

Pivoting a DataFrame means reshaping it by setting the index, columns, and values based on the existing columns. The pivot() method takes three main arguments: index, columns, and values.

For example, let us assume that you want to create a pivot table to display the average credit score for each geography and gender combination in the given dataset:

```python
# Calculate the average credit score for each combination of geography and gender
average_credit_score = data.groupby(['Geography', 'Gender'])['CreditScore'].mean().reset_index()


# Pivot the DataFrame
pivot_data = average_credit_score.pivot(index='Geography', columns='Gender', values='CreditScore')
```

The resulting pivot_data DataFrame will have 'Geography' as the index, 'Gender' as the columns, and the average credit score as the values.

Melting DataFrames using melt() Function

Melting a DataFrame means reshaping it from a wide format to a long format by moving multiple columns into a single column. The melt() function takes three main arguments: id_vars, value_vars, and var_name.

For example, let us melt the pivot_data DataFrame back into a long format:

```python
melted_data = pd.melt(pivot_data.reset_index(), id_vars='Geography', value_vars=['Female', 'Male'], var_name='Gender',
```

```
value_name='AverageCreditScore')
```

The resulting melted_data DataFrame will have a long format with 'Geography', 'Gender', and 'AverageCreditScore' columns.

In summary, pivoting and melting DataFrames in Pandas are techniques for reshaping your data without altering the actual data. Pivoting involves transforming data from a long format to a wide format using the pivot() method, while melting involves transforming data from a wide format to a long format using the melt() function. These techniques can help you better understand, visualize, and analyze your data by changing its structure.

Data Transformation Functions: apply(), map(), and applymap()

Pandas offers data transformation functions, including apply(), map(), and applymap(), which provide the ability to apply a customized function to individual elements, rows, or columns within a DataFrame or Series. These functions enable data manipulation and transformation in a highly adaptable and efficient manner. With these functions, data can be transformed and manipulated to meet specific needs. Apply() applies a function to each element or row/column of a DataFrame, while map() applies a function to each element of a Series. Applymap() applies a function to each element of a DataFrame. This range of functions provides a comprehensive toolkit to transform and manipulate data.

apply() Function

The apply() function is used to apply a function along an axis of the DataFrame (either rows or columns) or to a Series.

For example, let us assume that you want to calculate the range (max-min) for each numeric column in the data DataFrame:

```python
def col_range(col):
    return col.max() - col.min()


column_ranges = data.apply(col_range)
```

The apply() function can also be used with lambda functions. To achieve the same result as above:

```
column_ranges = data.apply(lambda col: col.max() - col.min())
```

map() Function

The map() function is used to apply a function to each element of a Series. It's useful when you want to perform element-wise transformations on a single column.

For example, let us assume that you want to update the 'Geography' column in the data DataFrame to have uppercase values:

```
data['Geography'] = data['Geography'].map(lambda x: x.upper())
```

applymap() Function

The applymap() function is used to apply a function to each element of a DataFrame. It's useful when you want to perform element-wise transformations on multiple columns.

For example, let us assume that you want to round all numeric columns in the data DataFrame to integers:

```
numeric_cols = data.select_dtypes(include='number').columns
data[numeric_cols] = data[numeric_cols].applymap(lambda x: round(x))
```

In summary, data transformation functions in Pandas like apply(), map(), and applymap() allow you to apply a specific function to elements, rows, or columns in a DataFrame or Series. The apply() function is used to apply a function along an axis of the DataFrame or to a Series, the map() function is used to apply a function to each element of a Series, and the applymap() function is used to apply a function to each element of a DataFrame. These functions help you transform and manipulate data in a flexible and efficient manner.

Grouping and Aggregating Data

In Pandas, the process of grouping and aggregating data involves utilizing the "groupby()" function to group data based on one or more columns, followed by the application of an aggregation function to the grouped data. This technique proves to be useful when analyzing data based on specific categories or combinations of categories. Through the "groupby()" function, it is possible to organize data into groups that have similar values and perform mathematical operations on the data within those groups. Grouping and aggregating data allows for a better understanding of patterns within the dataset and aids in identifying any trends or insights that may not have been noticeable otherwise.

Let us work through an example using the Customer_Churn_Modelling.csv dataset. We'll group the data by 'Geography' and 'Gender' and calculate the average credit score for each group.

Read the data

```python
import pandas as pd

data = pd.read_csv("https://raw.githubusercontent.com/kittenpub/database-repository/main/Customer_Churn_Modelling.csv")
```

Group the data using groupby()

```python
grouped_data = data.groupby(['Geography', 'Gender'])
```

This creates a groupby object with the data grouped by 'Geography' and 'Gender'. You can now apply aggregation functions to the grouped data.

Aggregate the data using an aggregation function

In this example, we'll use the mean() function to calculate the average credit score for each group:

```python
average_credit_scores = grouped_data['CreditScore'].mean()
```

The average_credit_scores variable now contains a Series with the average credit score for each combination of 'Geography' and 'Gender'.

You can also use the agg() function to apply multiple aggregation functions at once. For example, let us calculate the mean, minimum, and maximum credit scores for each group:

```python
credit_score_summary = grouped_data['CreditScore'].agg(['mean', 'min', 'max'])
```

The credit_score_summary variable now contains a DataFrame with the mean, minimum, and maximum credit scores for each combination of 'Geography' and 'Gender'.

In summary, grouping and aggregating data in Pandas involve using the groupby() function to group the data based on one or more columns and then applying an aggregation function to the grouped data. This is useful for analyzing your data by specific categories or combinations of categories. In the given example, we grouped the data by 'Geography' and 'Gender' and calculated the average credit score for each group.

Custom Aggregation Functions

Pandas provide the flexibility to define and utilize custom aggregation functions that can be applied to grouped data. This feature comes in handy when the built-in aggregation functions are unable to cater to the specific needs of the user, or when the user desires to conduct intricate calculations on the data. By defining custom aggregation functions, the user can tailor their analysis to suit their requirements and derive meaningful insights from their data. This functionality enhances the versatility of Pandas and empowers users to leverage the power of Python to manipulate data effectively. Whether it's a simple calculation or a complex operation, custom aggregation functions enable the user to process their data efficiently and derive maximum value from it.

Let us work through an example using the Customer_Churn_Modelling.csv dataset. We'll group the data by 'Geography' and 'Gender' and calculate the range (max-min) of the credit scores for each group using a custom aggregation function.

Read the data

```
import pandas as pd

data = pd.read_csv("https://raw.githubusercontent.com/kittenpub/database-repository/main/Customer_Churn_Modelling.csv")
```

Define the custom aggregation function
In this example, we'll create a function to calculate the range (max-min) of a given Series:

```
def credit_score_range(series):
    return series.max() - series.min()
```

Group the data using groupby()

```
grouped_data = data.groupby(['Geography', 'Gender'])
```

This creates a groupby object with the data grouped by 'Geography' and 'Gender'. You can now apply your custom aggregation function to the grouped data.

Apply the custom aggregation function
In this example, we'll use the agg() function to apply our custom aggregation function (credit_score_range) to the 'CreditScore' column of each group:

```
credit_score_ranges = grouped_data['CreditScore'].agg(credit_score_range)
```

The credit_score_ranges variable now contains a Series with the range of credit scores for each combination of 'Geography' and 'Gender'.

You can also apply multiple aggregation functions at once, including built-in and custom functions. For example, let us calculate the mean and range of credit scores for each group:

```
credit_score_summary = grouped_data['CreditScore'].agg(['mean',
credit_score_range])
```

The credit_score_summary variable now contains a DataFrame with the mean and range of credit scores for each combination of 'Geography' and 'Gender'.

In summary, custom aggregation functions in Pandas allow you to create your own aggregation functions and apply them to grouped data. This is useful when built-in aggregation functions don't meet your specific needs, or when you want to perform more complex calculations on your data. In the given example, we created a custom aggregation function to calculate the range of credit scores and applied it to the data grouped by 'Geography' and 'Gender'.

Summary

In this chapter, we discussed various advanced operations and techniques in Pandas as below:

Grouping and Aggregating Data: We explained how to use the groupby() function to group data based on one or more columns and apply aggregation functions to the grouped data. We used the Customer_Churn_Modelling.csv dataset as an example and grouped the data by 'Geography' and 'Gender' to calculate the average credit score for each group.

Custom Aggregation Functions: We taught how to create custom aggregation functions and apply them to grouped data. We used the same dataset and created a custom aggregation function to calculate the range (max-min) of credit scores for each group based on 'Geography' and 'Gender'.

Data Transformation Functions: We described the apply(), map(), and applymap() functions in Pandas, which allow you to apply specific functions to elements, rows, or columns in a DataFrame or Series. We provided examples of using these functions on the given dataset for various tasks, such as calculating the range of values in numeric columns, updating the 'Geography' column to uppercase values, and rounding all numeric columns to integers.

Merging and Joining DataFrames: We explained the concepts of merging and joining DataFrames and provided a practical example of how to perform these operations on the given dataset.

Concatenating and Appending DataFrames: We explained the concepts of concatenating and appending DataFrames and provided practical examples of how to perform these operations on the given dataset.

Pivoting and Melting DataFrames: We described the concepts of pivoting and melting DataFrames and provided practical examples of how to perform these operations on the given dataset.

Throughout this chapter, we focused on practical examples and step-by-step instructions to help you learn and apply these advanced Pandas techniques to the Customer_Churn_Modelling.csv dataset.

CHAPTER 5: TIME SERIES AND DATETIME OPERATIONS

Introduction to Time Series Data

Time series data is a sequence of data points collected or recorded in a chronological order at regular intervals of time. These data points can be observations or measurements of a variable of interest, such as stock prices, temperature readings, or sales figures. The primary characteristic of time series data is its inherent temporal structure, which provides insights into patterns, trends, seasonality, and the relationships between different variables over time.

Analyzing time series data allows us to understand past behavior, identify the factors influencing the variable, and forecast future values. Time series analysis has numerous applications across various domains, such as finance, economics, meteorology, healthcare, and social sciences. Pandas, a widely-used data manipulation library in Python, has become a go-to tool for time series data analysis because of its extensive support for handling and processing temporal data.

Some of the key features of Pandas that make it a powerful tool for time series analysis include:

- Data structures designed for time series data: Pandas provides the Series and DataFrame data structures, which have built-in support for handling time-based indices. This enables efficient and flexible manipulation of time series data.
- Time-based indexing and slicing: Pandas supports time-based indexing, which makes it easy to select, filter, and modify data based on date and time values. You can perform range-based queries, select specific periods, or resample the data at different frequencies using simple and intuitive syntax.
- Date and time objects: Pandas integrates with Python's built-in datetime module and provides additional functionality through its own Timestamp, Period, and Timedelta objects. These objects simplify working with dates, times, and intervals, allowing you to perform calculations, comparisons, and conversions with ease.
- Resampling and frequency conversion: Pandas provides powerful resampling methods that enable you to change the frequency of your data, aggregate data into different time periods, or interpolate missing values. This is useful when you want to analyze your data at different levels of granularity or fill in gaps in your dataset.
- Time zone handling: Time series data often comes from sources with different time zones, and managing these time zones can be a complex task. Pandas offers robust support for time zone conversion, localization, and normalization, making it easy to work with data from multiple time zones.
- Rolling and expanding window functions: Pandas allows you to apply rolling and expanding window functions to time series data, which can help you identify trends, compute moving averages, or calculate cumulative sums. These functions can be customized with various window types and aggregation functions to suit your

needs.

- Shifting and lagging data: Time series analysis often involves comparing data points across different time periods or calculating the differences between consecutive values. Pandas provides methods for shifting and lagging data, enabling you to perform these operations efficiently and accurately.
- Integration with other libraries: Pandas seamlessly integrates with other Python libraries, such as NumPy, SciPy, and scikit-learn, allowing you to perform advanced statistical analysis, modeling, and machine learning on your time series data.
- Input/Output support for various file formats: Pandas supports reading and writing data from a wide range of file formats, including CSV, Excel, JSON, and SQL databases. This makes it easy to import and export time series data from different sources and formats.

In summary, time series data is a sequence of data points collected or recorded in chronological order at regular intervals of time. Analyzing time series data provides valuable insights into patterns, trends, and relationships over time. Pandas, a powerful data manipulation library in Python, has become a go-to tool for time series data analysis due to its extensive support for handling and processing temporal data

DateTime Objects and Functions

DateTime objects and functions in Pandas are designed to handle and manipulate date and time data efficiently. The library extends the functionality of Python's built-in datetime module, providing additional support through its Timestamp, Period, and Timedelta objects. These tools enable users to easily manipulate and manage time-based data, making it an essential tool for many data analysis and modeling tasks. By leveraging Pandas' powerful capabilities, users can streamline their workflow and gain insights into temporal data with ease. We'll discuss these objects and functions in detail and demonstrate how to use them with the Customer_Churn_Modelling.csv dataset.

Timestamp

A Timestamp object represents a single point in time. It is the Pandas equivalent of Python's datetime.datetime object. You can create a Timestamp object by passing a date or datetime string to the pd.Timestamp() function.

```python
import pandas as pd


timestamp_example = pd.Timestamp('2023-04-12 12:00:00')
```

```python
print(timestamp_example)
```

Period

A Period object represents a time span or duration, such as a day, a month, or a year. You can create a Period object by passing a date or datetime string and a frequency to the pd.Period() function.

```python
period_example = pd.Period('2023-04', freq='M')
print(period_example)
```

Timedelta

A Timedelta object represents the difference between two dates or times. It is the Pandas equivalent of Python's datetime.timedelta object. You can create a Timedelta object by passing a duration string to the pd.Timedelta() function.

```python
timedelta_example = pd.Timedelta('1 day')
print(timedelta_example)
```

Now let us see how to use these DateTime objects and functions with the Customer_Churn_Modelling.csv dataset. First, we'll load the dataset:

```python
data = pd.read_csv("https://raw.githubusercontent.com/kittenpub/database-repository/main/Customer_Churn_Modelling.csv")
```

Assuming the dataset has a 'Date' column in the format 'YYYY-MM-DD', we can convert this column to a datetime dtype using the pd.to_datetime() function:

```python
data['Date'] = pd.to_datetime(data['Date'])
```

Now the 'Date' column contains Timestamp objects that can be used for time-based indexing, slicing, and other operations.

For example, we can filter the data for a specific date range:

```python
date_range_mask = (data['Date'] >= '2020-01-01') & (data['Date'] <= '2020-12-31')
filtered_data = data[date_range_mask]
```

We can also calculate the difference between two dates in the dataset:

```python
date_difference = data['Date'].max() - data['Date'].min()
print(date_difference)
```

Finally, we can use DateTime objects and functions to create new columns based on existing date columns. For example, let us add a column that represents the month and year of each date in the 'Date' column:

```python
data['Month_Year'] = data['Date'].apply(lambda x: pd.Period(x, freq='M'))
```

In summary, DateTime objects and functions in Pandas, such as Timestamp, Period, and Timedelta, provide a powerful and flexible way to handle date and time data. By using these objects and functions with the Customer_Churn_Modelling.csv dataset, we can easily manipulate and analyze time series data for various tasks, such as filtering, indexing, and creating new columns based on existing date columns.

Time Series Data Manipulation

Time series data manipulation involves processing and transforming time-based data to facilitate analysis and gain insights. In this section, we will go through some practical examples of time series data manipulation using the Pandas library and the Customer_Churn_Modelling.csv dataset. First, let us load the dataset and convert the 'Date' column to a datetime dtype:

```python
import pandas as pd

data = pd.read_csv("https://raw.githubusercontent.com/kittenpub/database-repository/main/Customer_Churn_Modelling.csv")
data['Date'] = pd.to_datetime(data['Date'])
```

Assuming the 'Date' column in the dataset represents the date when customer data was recorded, we can perform various time series data manipulation tasks:

Set the date column as the index

```python
data.set_index('Date', inplace=True)
```

Select data for a specific time range

```python
selected_data = data['2020-01-01':'2020-12-31']
```

Resample data to a different frequency

Resample the data to monthly frequency and calculate the mean:

```python
monthly_data = data.resample('M').mean()
```

Resample the data to quarterly frequency and calculate the sum:

```python
quarterly_data = data.resample('Q').sum()
```

Calculate rolling statistics

Calculate a 7-day rolling mean for the 'CreditScore' column:

```python
data['CreditScore_7D_Rolling_Mean'] =
data['CreditScore'].rolling(window=7).mean()
```

Calculate a 30-day rolling standard deviation for the 'CreditScore' column:

```python
data['CreditScore_30D_Rolling_Std'] =
data['CreditScore'].rolling(window=30).std()
```

Calculate the difference between consecutive values (lagging):

```python
data['CreditScore_Diff'] = data['CreditScore'].diff()
```

Shift the data forward or backward in time

Shift the 'CreditScore' column one period forward:

```python
data['CreditScore_Shifted_Forward'] = data['CreditScore'].shift(1)
```

Shift the 'CreditScore' column one period backward:

```python
data['CreditScore_Shifted_Backward'] = data['CreditScore'].shift(-1)
```

These are just a few examples of time series data manipulation tasks that can be performed using Pandas. By applying these techniques to the Customer_Churn_Modelling.csv dataset, you can analyze customer data over time, identify trends and patterns, and gain valuable insights into customer behavior and churn.

Frequency Conversion and Resampling

In time series analysis, frequency conversion and resampling are common techniques used to modify the frequency of data by either combining or estimating values between original data points. With Pandas, these operations can be executed efficiently and with great flexibility. Pandas offers a variety of powerful tools to handle frequency conversion and resampling in time series data, which can help to better understand and analyze trends and patterns in the data.

In this section, we'll demonstrate how to perform frequency conversion and resampling using the Customer_Churn_Modelling.csv dataset.

First, let us load the dataset and convert the 'Date' column to a datetime dtype:

```python
import pandas as pd

data = pd.read_csv("https://raw.githubusercontent.com/kittenpub/database-repository/main/Customer_Churn_Modelling.csv")
data['Date'] = pd.to_datetime(data['Date'])
```

Next, we'll set the 'Date' column as the index:

```
data.set_index('Date', inplace=True)
```

Now let us perform frequency conversion and resampling on the dataset:

Resampling

This involves changing the frequency of the data by aggregating data points. The resample() method in Pandas provides an easy way to perform resampling.

For example, let us resample the data to a monthly frequency and calculate the mean of the 'CreditScore' column:

```
monthly_data_mean = data['CreditScore'].resample('M').mean()
```

Similarly, we can resample the data to a quarterly frequency and calculate the sum of the 'CreditScore' column:

```
quarterly_data_sum = data['CreditScore'].resample('Q').sum()
```

Frequency Conversion

This involves changing the frequency of the data without aggregating or interpolating data points. The asfreq() method in Pandas can be used for frequency conversion. For example, let us convert the daily frequency of the dataset to a weekly frequency:

```
weekly_data = data.asfreq('W')
```

Keep in mind that when you perform frequency conversion using asfreq(), you may end up with missing values (NaN) if the new frequency is less frequent than the original frequency.

By using frequency conversion and resampling techniques with the Customer_Churn_Modelling.csv dataset, you can analyze the data at different time frequencies, identify trends and patterns, and gain insights into customer behavior and churn. These techniques allow you to perform more advanced time series analysis and make better decisions based on the data.

Time Zone Handling

Time zone handling refers to the management of time-related information that originates from different geographical areas, involving the conversion of time between time zones and the execution of various time-based computations while accounting for differences in time zones. It involves the ability to accurately interpret and manipulate date and time data to ensure that the correct time is displayed, recorded, and used in various applications. This process is crucial for industries such as travel, finance, and communication, where time is a critical factor in decision-making and the accuracy of information is paramount.

In this section, we'll demonstrate how to perform time zone handling using the Pandas library and the Customer_Churn_Modelling.csv dataset.

Convert to Datetime Dtype

First, let us load the dataset and convert the 'Date' column to a datetime dtype:

```
import pandas as pd

data = pd.read_csv("https://raw.githubusercontent.com/kittenpub/database-repository/main/Customer_Churn_Modelling.csv")
data['Date'] = pd.to_datetime(data['Date'])
```

Set Date Column as Index

Next, we'll set the 'Date' column as the index:

```
data.set_index('Date', inplace=True)
```

Now let us explore time zone handling with the dataset:

By default, the datetime index is timezone-naive. We can localize it to a specific time zone using the tz_localize() method. For example, let us localize the index to the 'US/Eastern' time zone:

```
data_tz = data.tz_localize('US/Eastern')
```

Convert Datetime Index to Different Time Zone

Once the datetime index is localized, we can convert it to a different time zone using the tz_convert() method. For example, let us convert the 'US/Eastern' time zone index to the 'Europe/London' time zone:

```
data_tz_london = data_tz.tz_convert('Europe/London')
```

Perform Time-based Calculations with Time Zone-aware Data

Time zone-aware data can be used for various time-based calculations, such as calculating the time difference between two dates. For example, let us calculate the time difference between the maximum and minimum dates in the dataset, taking into account the 'Europe/London' time zone:

```
time_difference_london = data_tz_london.index.max() -
data_tz_london.index.min()
print(time_difference_london)
```

By performing time zone handling with the Customer_Churn_Modelling.csv dataset, you can ensure that your analysis takes into account time zone differences and provides accurate insights into customer behavior and churn. This is particularly important when dealing with data from different regions or when working with daylight saving time changes, as time zone handling can help you avoid errors and inconsistencies in your analysis.

Periods and Period Arithmetic

In the Pandas library, working with time-related data involves using Periods and Period Arithmetic, which differ from working with specific points in time. Instead, Periods represent a time span, such as a day, month, or year, allowing for more flexible analysis of time-based data. With Period Arithmetic, it is possible to perform calculations and comparisons between these time spans, enabling users to analyze trends and patterns over time more efficiently. For example, one can perform arithmetic operations such as adding or subtracting periods, determining the difference between two periods, or comparing whether two periods overlap. By utilizing Periods and Period Arithmetic in Pandas, analysts can gain a better understanding of temporal data and make more informed decisions based on trends and patterns over extended periods.

First, let us load the dataset and convert the 'Date' column to a datetime dtype:

```python
import pandas as pd

data = pd.read_csv("https://raw.githubusercontent.com/kittenpub/database-repository/main/Customer_Churn_Modelling.csv")
data['Date'] = pd.to_datetime(data['Date'])
```

Next, we'll set the 'Date' column as the index:

```python
data.set_index('Date', inplace=True)
```

Now let us explore Periods and Period Arithmetic with the dataset:

Create Period Object

A period object represents a time span. Let us create a period object representing the month of January 2020:

```python
january_2020 = pd.Period('2020-01', freq='M')
```

Perform Period Arithmetic

You can perform arithmetic operations with period objects, such as adding or subtracting time spans. For example, let us add one month to the january_2020 period object:

```python
february_2020 = january_2020 + 1
```

Convert Datetime Index to PeriodIndex

To work with periods in the dataframe, you can convert the datetime index to a PeriodIndex. For example, let us convert the daily datetime index to a monthly PeriodIndex:

```python
data_period = data.to_period(freq='M')
```

Perform Calculations with PeriodIndex

You can perform various calculations using a PeriodIndex. For example, let us calculate the number of unique months in the dataset:

```
unique_months = len(data_period.index.unique())
```

Group Data by Periods

PeriodIndex can be used to group data based on specific time spans. For example, let us group the data by month and calculate the mean of the 'CreditScore' column:

```
monthly_data_mean = data_period.groupby(level=0)['CreditScore'].mean()
```

By working with Periods and Period Arithmetic in the Customer_Churn_Modelling.csv dataset, you can analyze time-based data in terms of time spans instead of specific points in time. This allows you to perform more advanced time series analysis and gain insights into customer behavior and churn based on various time spans, such as months, quarters, or years.

Advanced Time Series Techniques

Pandas offers several advanced time series techniques that can help you analyze and process time series data more effectively. Let us explore some of these techniques and how they can be applied to the Customer_Churn_Modelling.csv dataset.

First, let us load the dataset and convert the 'Date' column to a datetime dtype:

```
import pandas as pd

data = pd.read_csv("https://raw.githubusercontent.com/kittenpub/database-repository/main/Customer_Churn_Modelling.csv")
data['Date'] = pd.to_datetime(data['Date'])
```

Next, we'll set the 'Date' column as the index:

```
data.set_index('Date', inplace=True)
```

Now, let us explore some advanced time series techniques:

Rolling Windows

Rolling windows can be used to calculate various statistics, such as the mean, median, or standard deviation, over a moving window of data points. This is useful for smoothing out noisy data or identifying trends in the data.

```python
# Calculate the rolling mean of the 'CreditScore' column with a window size of 7 days
data['CreditScore_rolling_mean'] = data['CreditScore'].rolling(window='7D').mean()
```

Exponential Moving Average

The exponential moving average (EMA) is a type of weighted moving average that gives more importance to recent data points. This can help identify trends more effectively than a simple moving average.

```python
# Calculate the EMA of the 'CreditScore' column with a span of 7 days
data['CreditScore_ema'] = data['CreditScore'].ewm(span=7).mean()
```

Time Series Decomposition

Time series decomposition involves breaking down a time series into its components, such as trend, seasonality, and residual (noise). This can help you better understand the underlying patterns in the data.

```python
from statsmodels.tsa.seasonal import seasonal_decompose

# Decompose the 'CreditScore' time series data
decomposition = seasonal_decompose(data['CreditScore'], period=30, model='additive')

# Plot the decomposition components
decomposition.plot()
```

Time Series Forecasting

Time series forecasting involves predicting future values of a time series based on historical data. There are various forecasting techniques, such as autoregression (AR), moving average (MA), and ARIMA (AutoRegressive Integrated Moving Average).

```python
from statsmodels.tsa.arima.model import ARIMA

# Fit an ARIMA model to the 'CreditScore' time series data
model = ARIMA(data['CreditScore'], order=(1, 1, 1))
model_fit = model.fit()

# Predict the next 5 days of 'CreditScore' values
predictions = model_fit.forecast(steps=5)
```

By using these advanced time series techniques with the Customer_Churn_Modelling.csv dataset, you can gain deeper insights into customer behavior and churn, identify trends and patterns, and even make predictions about future customer behavior. These techniques can help you make more informed decisions based on your analysis of the data.

Summary

In this chapter, we explored the concept of time series data and how the pandas library is a powerful tool for analyzing time series datasets. We also looked at DateTime objects and functions in Pandas 2.0, including how to use them to manipulate time series data. We covered time series data manipulation, frequency conversion and resampling, time zone handling, periods and period arithmetic, and advanced time series techniques.

We began by loading the Customer_Churn_Modelling.csv dataset and converting the 'Date' column to a datetime dtype, followed by setting the 'Date' column as the index. We then explored different aspects of time series data analysis in pandas, including rolling windows, exponential moving averages, time series decomposition, and time series forecasting.

We first covered rolling windows and how they can be used to calculate statistics over a moving window of data points. This technique is useful for smoothing out noisy data or identifying trends in the data. We demonstrated how to calculate the rolling mean of the 'CreditScore' column with a window size of 7 days.

Next, we explored exponential moving averages, which are a type of weighted moving average that give more importance to recent data points. This technique can help identify trends more effectively than a simple moving average. We showed how to calculate the EMA of the 'CreditScore' column with a span of 7 days.

We then covered time series decomposition, which involves breaking down a time series into its components, such as trend, seasonality, and residual (noise). This technique can help you better understand the underlying patterns in the data. We demonstrated how to decompose the 'CreditScore' time series data and plot the decomposition components.

Finally, we looked at time series forecasting, which involves predicting future values of a time series based on historical data. We covered various forecasting techniques, such as autoregression (AR), moving average (MA), and ARIMA (AutoRegressive Integrated Moving Average). We demonstrated how to fit an ARIMA model to the 'CreditScore' time series data and predict the next 5 days of 'CreditScore' values.

By using these advanced time series techniques with the Customer_Churn_Modelling.csv dataset, you can gain deeper insights into customer behavior and churn, identify trends and patterns, and even make predictions about future customer behavior. These techniques can help you make more informed decisions based on your analysis of the data.

CHAPTER 6:
PERFORMANCE OPTIMIZATION AND SCALING

Memory and Computation Efficiency

Memory and computation efficiency are crucial aspects of data analysis and processing, particularly when working with large datasets. These concepts aim to minimize the resources consumed by a system or application, optimizing performance, and allowing for faster and more scalable solutions. Understanding the importance of memory and computation efficiency can help you develop more efficient and robust applications.

The following are some key reasons why memory and computation efficiency are important:

- Resource limitations: Computers have finite resources, such as RAM and CPU processing power. Efficient memory and computation usage allow you to work within these constraints, preventing your system from crashing or slowing down. This is especially important when working with large datasets, which can easily consume all available resources if not managed efficiently.
- Scalability: Efficient memory and computation management enable your applications to scale more effectively, allowing you to process larger datasets or handle more requests simultaneously. This is particularly important in the age of big data, where the volume of data being generated and processed is continually increasing.
- Faster processing: Optimizing memory and computation usage can significantly speed up your data processing and analysis tasks. Faster processing times can lead to quicker insights and decision-making, which can be critical in many industries and applications.
- Cost savings: Efficient use of resources can lead to cost savings, especially in cloud computing environments where resources are billed based on usage. By optimizing memory and computation, you can reduce your infrastructure costs and make your applications more cost-effective.
- Better user experience: Applications that are optimized for memory and computation efficiency often have better performance, leading to improved user experience. Users will typically have a more positive experience with applications that load faster, respond quickly to user inputs, and can handle large amounts of data without crashing or slowing down.
- Environmental impact: Efficient computing helps reduce energy consumption, which has a positive impact on the environment. As data processing demands continue to grow, it's essential to find ways to minimize the energy footprint of computing tasks.

Memory and computation efficiency play a vital role in data analysis and processing. By optimizing your applications for efficient memory and computation usage, you can achieve better performance, scalability, cost savings, and environmental benefits, ultimately leading to a more robust and effective solution.

For the given database on Customer Churn Modeling available at below:
https://github.com/kittenpub/database-repository/blob/main/Customer_Churn_Modelling.csv
the memory and computation efficiency can be applied in the following ways:

Choose Appropriate Data types

Choose appropriate data types for each column in the DataFrame. Pandas automatically detects the data types, but sometimes it's better to specify them manually, especially for large datasets. For example, you can use 'category' data type for columns with a limited number of unique values.

```python
import pandas as pd
dtypes = {'Geography': 'category', 'Gender': 'category'}
df = pd.read_csv('Customer_Churn_Modelling.csv', dtype=dtypes)
```

Loading in Chunks

If the dataset is very large, load it in chunks instead of reading it all at once. This helps to manage memory usage and perform data manipulation operations in smaller, more manageable pieces.

```python
chunksize = 1000
chunks = []
for chunk in pd.read_csv('Customer_Churn_Modelling.csv',
chunksize=chunksize):
    # Perform any data processing or filtering on the chunk
    chunks.append(chunk)
df = pd.concat(chunks, axis=0)
```

Use Built-in Optimized Functions

Pandas provides a variety of vectorized functions and methods that are optimized for performance. Use these functions instead of custom or less-efficient alternatives.

Parallel Processing

If you have a multicore processor, you can use parallel processing to speed up computations on large datasets. This can be done using libraries like Dask or by parallelizing your code using the 'multiprocessing' module.

Use In-place Operations

When possible, use in-place operations to avoid creating unnecessary copies of your data, which can consume additional memory.

```
df.drop(columns=['Unwanted_Column'], inplace=True)
```

By applying these memory and computation efficiency techniques, you can ensure that your data analysis tasks on the given dataset will be more efficient and consume fewer resources, making it easier to work with large datasets and perform complex operations.

Utilizing Dask for Parallel and Distributed Computing

Dask is a powerful library that enables parallel and distributed computing in Python, making it easier to work with large datasets that may not fit in memory. Dask extends the capabilities of Pandas by providing parallelized versions of common Pandas data structures, such as Dask DataFrame, which is a large parallel DataFrame composed of smaller Pandas DataFrames.

The given below demonstrates how you can use Dask to perform parallel and distributed computing with Pandas:

Installing Dask

First, you need to install Dask. You can do this using pip or conda:

```
pip install dask[complete]
```

Importing Dask

After installing Dask, you can import it into your Python script or notebook:

```python
import dask.dataframe as dd
```

Reading Data

Read your dataset using Dask instead of Pandas. Dask provides similar functions to Pandas for reading data, such as read_csv. By using Dask to read the data, you can efficiently work with larger-than-memory datasets:

```python
ddf = dd.read_csv('large_dataset.csv')
```

Manipulating Data

Perform data manipulation tasks using Dask's API, which is similar to Pandas. Dask will automatically parallelize these operations:

```python
# Filtering rows
filtered_ddf = ddf[ddf['column_name'] > value]

# Grouping and aggregation
grouped_ddf = ddf.groupby('group_column').agg({'value_column': 'mean'})
```

Computing Result

Since Dask operations are lazy by default, you need to call the compute() method to execute the operations and return the result as a Pandas DataFrame:

```python
result_df = grouped_ddf.compute()
```

Distributed Computing

To use Dask for distributed computing, you can set up a Dask cluster using Dask's distributed scheduler. First, install the dask.distributed library, and then create a Dask client:

```python
from dask.distributed import Client

client = Client()  # This will create a local cluster by default
```

After creating a Dask client, any operations you perform using Dask will automatically be distributed across the available workers in the cluster.

In summary, Dask is a powerful library that can help you parallelize and distribute your Pandas computations. By using Dask's API, which is similar to Pandas, you can efficiently work with large datasets and perform complex data analysis tasks more quickly and with fewer resources.

Querying and Filtering Data Efficiently

To perform querying and filtering data efficiently on the Customer_Churn_Modelling database, you can use a combination of Pandas and Dask for memory and computation efficiency. The given below demonstrates an example of how to filter and query the data effectively:

First, you need to import the necessary libraries and read the dataset:

```python
import pandas as pd
import dask.dataframe as dd


# Read the dataset using Dask
ddf = dd.read_csv('https://raw.githubusercontent.com/kittenpub/database-repository/main/Customer_Churn_Modelling.csv')
```

Now, let us assume you want to filter the data based on specific criteria, such as customers with a credit score above 700 and who are active members:

```python
# Filter data based on conditions
filtered_ddf = ddf[(ddf['CreditScore'] > 700) & (ddf['IsActiveMember'] == 1)]
```

Next, you might want to query the data to find the average balance for each unique geography:

```python
# Group by geography and calculate the average balance
average_balance_by_geography = filtered_ddf.groupby('Geography')['Balance'].mean()
```

Since Dask operations are lazy by default, you need to call the compute() method to execute the operations and get the result as a Pandas DataFrame:

```
result_df = average_balance_by_geography.compute()
```

Now, you have a Pandas DataFrame with the average balance for each unique geography based on the filtered data.

By using Dask to perform these filtering and querying tasks, you can efficiently work with large datasets and perform complex operations more quickly. Additionally, you can further optimize the performance by using appropriate data types, loading the data in chunks, and parallelizing the computations as described in the previous section.

Vectorized Operations and Performance

Vectorized operations refer to performing operations on entire arrays or data structures rather than on individual elements in a loop. In the context of Pandas, vectorized operations are applied to Series and DataFrames using NumPy, which is an underlying library that provides efficient array operations. These operations are usually faster than using traditional loops because they are optimized and executed in compiled code.

The benefits of vectorized operations include:
- Improved performance: Vectorized operations are generally faster than using loops because they utilize low-level optimizations and eliminate the overhead of Python loops.
- Better readability: Vectorized code is often more concise and easier to read compared to equivalent loop-based code.
- Easier code maintenance: Since vectorized code is more concise, it is generally easier to maintain and less prone to errors.

Performing Vectorized Operations

The given below demonstrates how you can perform vectorized operations on the given Customer_Churn_Modelling database:

First, read the dataset using Pandas:

```
import pandas as pd
```

```
url = 'https://raw.githubusercontent.com/kittenpub/database-
repository/main/Customer_Churn_Modelling.csv'
df = pd.read_csv(url)
```

Suppose you want to calculate the ratio of balance to the estimated salary for each customer. You can perform a vectorized operation by dividing the 'Balance' column by the 'EstimatedSalary' column:

```
df['BalanceToSalaryRatio'] = df['Balance'] / df['EstimatedSalary']
```

Here, the division operation is applied element-wise to each corresponding element in the 'Balance' and 'EstimatedSalary' columns, creating a new 'BalanceToSalaryRatio' column in the DataFrame.

Another example is applying a function to a column in a vectorized manner. Suppose you want to categorize customers based on their age: 'Young' for customers below 30, 'Adult' for customers between 30 and 60, and 'Senior' for customers above 60. You can use the pd.cut() function to perform this categorization:

```
age_bins = [0, 30, 60, float('inf')]
age_labels = ['Young', 'Adult', 'Senior']
df['AgeCategory'] = pd.cut(df['Age'], bins=age_bins, labels=age_labels)
```

In this case, the pd.cut() function is applied to the 'Age' column, creating a new 'AgeCategory' column with the appropriate age categories.

By using vectorized operations in your data processing and analysis tasks, you can improve performance, readability, and maintainability of your code. It's important to utilize these operations whenever possible to take full advantage of the capabilities provided by Pandas and NumPy.

Using Cython and Numba for Speed

In data analysis operations, speed issues can arise from several factors, including:
- Looping over large datasets: Iterating over large datasets using Python loops can be slow due to the interpreted nature of Python and the overhead associated with loops.

- Inefficient implementations: Using non-vectorized or non-optimized implementations for operations on large datasets can lead to slow processing times.
- Complex calculations: Performing complex calculations on large datasets can be computationally expensive and slow, particularly if the calculations cannot be easily parallelized or optimized.

Cython and Numba are two popular solutions for speeding up data analysis operations in Python:

Cython

Cython is a superset of Python that allows you to write C extensions for Python in a Python-like syntax. It lets you statically type variables and functions, which can lead to significant performance improvements when compiled to C. Cython is particularly useful when you have a performance-critical section of your code that can benefit from the speed of C.

Numba

Numba is a Just-In-Time (JIT) compiler for Python that translates a subset of Python and NumPy code into machine code at runtime. Numba enables you to write high-performance functions in Python by using decorators to specify that a function should be compiled by Numba. Numba is especially helpful when you have computationally expensive functions that cannot be easily vectorized using NumPy or Pandas.

Install Cython and Numba

To install Cython and Numba, use pip:

```
pip install cython numba
```

Now, let us see how Cython and Numba can be used to speed up the performance of a simple example. Consider the following Python function that computes the sum of squares for a range of numbers:

```python
def sum_of_squares(n):
    total = 0
    for i in range(n):
        total += i ** 2
```

```
    return total
```

Using Cython to Speed Up Function

First, you need to load the Cython extension in your Jupyter Notebook:

```
%load_ext cython
```

Next, rewrite the function using Cython by adding the %%cython magic command and specifying the types of the variables:

```
%%cython
def cython_sum_of_squares(int n):
    cdef int total = 0
    cdef int i
    for i in range(n):
        total += i ** 2
    return total
```

Using Numba to Speed Up Function

Import Numba and use the @jit decorator to indicate that the function should be compiled with Numba:

```
from numba import jit

@jit(nopython=True)
def numba_sum_of_squares(n):
    total = 0
    for i in range(n):
        total += i ** 2
    return total
```

Compare Performance

Now, you can compare the performance of the original function, the Cython-optimized version, and the Numba-optimized version:

```python
import timeit

n = 10_000

print("Python function:")
%timeit sum_of_squares(n)

print("\nCython function:")
%timeit cython_sum_of_squares(n)

print("\nNumba function:")
%timeit numba_sum_of_squares(n)
```

You should see that both the Cython and Numba implementations are significantly faster than the original Python function.

In summary, Cython and Numba are powerful tools for optimizing the performance of your data analysis operations in Python. By using these tools, you can significantly speed up your code and improve the efficiency of your data processing tasks. Choose between Cython and Numba based on your specific requirements, the nature of the code, and the type of optimizations needed.

Debugging and Profiling Performance Issues

The given below is a more detailed elaboration of the common performance issues and their resolutions when using Pandas:

Memory Usage

Loading data in chunks

When reading data from a file or database, you can use the chunksize parameter to read the data in smaller chunks. This helps reduce memory usage and allows you to process the data incrementally.

```python
import pandas as pd

chunksize = 10 ** 5  # Adjust the chunk size based on your system's memory
for chunk in pd.read_csv("large_dataset.csv", chunksize=chunksize):
    # Perform your operations on each chunk
```

Dropping unnecessary columns or rows
It's a good practice to drop any columns or rows that you don't need in your analysis as early as possible. This reduces memory usage and speeds up subsequent operations.

```python
# Drop unnecessary columns
df = df.drop(['unnecessary_column1', 'unnecessary_column2'], axis=1)

# Drop unnecessary rows based on a condition
df = df[df['useful_column'] > threshold]
```

Using appropriate data types
Pandas automatically infers the data type of each column when reading data, which can sometimes lead to inefficient memory usage. You can reduce memory usage by specifying the correct data types or converting columns to more efficient types.

```python
# Specify data types when reading data
dtypes = {'column1': 'int32', 'column2': 'category'}
df = pd.read_csv('dataset.csv', dtype=dtypes)

# Convert existing columns to more efficient data types
df['column1'] = df['column1'].astype('int32')
df['column2'] = df['column2'].astype('category')
```

Non-vectorized Operations

Vectorized Pandas or NumPy operations

Whenever possible, use vectorized operations provided by Pandas and NumPy to perform calculations on entire columns or DataFrames at once. These operations are optimized and significantly faster than looping through rows.

```python
# Example of vectorized addition
df['sum_column'] = df['column1'] + df['column2']
```

Using apply()

If an operation cannot be easily vectorized, you can use the apply() method to apply a custom function along a specific axis (rows or columns) of a DataFrame.

```python
def custom_function(row):
    # Perform calculations on row elements
    return result

df['new_column'] = df.apply(custom_function, axis=1)
```

Using Cython or Numba

If performance is still an issue with non-vectorized operations, consider using Cython or Numba to optimize your custom functions. Both tools can provide significant speed improvements by compiling your Python code to native machine code or C.

Inefficient Chaining of Operations

Using the inplace parameter

Many Pandas methods provide an inplace parameter, which allows you to modify the DataFrame in place without creating a new object.

```python
df.drop(['column1', 'column2'], axis=1, inplace=True)
```

Using query() or eval()

Pandas provides query() and eval() methods that allow you to perform multiple operations in a single step, reducing the number of intermediate objects created. These methods can be faster and more memory-efficient than chaining operations.

```python
# Filter and calculate a new column in one step using query()
df = df.query('column1 > 100 and
    column2 < 200')
df['new_column'] = df['column1'] * df['column2']


# Perform multiple calculations in one step using eval()
df['new_column'] = df.eval('column1 * column2')
```

Slow Groupby and Aggregation Operations

Using `sort=False`

By default, Pandas sorts the groups when performing a `groupby()` operation, which can slow down the process. You can disable sorting by setting `sort=False`.

```python
grouped = df.groupby('column1', sort=False)
```

Using the `agg()` method

Instead of chaining multiple aggregation operations, you can use the `agg()` method to perform multiple aggregations in one step.

```python
result = grouped.agg({
    'column2': ['sum', 'mean'],
    'column3': ['max', 'min']
})
```

Slow Reading and Writing Operations

Using `compression`

When reading or writing data, you can use the `compression` parameter to specify a compression algorithm, reducing file size and I/O time.

```python
df = pd.read_csv('compressed_data.csv.gz', compression='gzip')
df.to_csv('compressed_output.csv.gz', compression='gzip')
```

Binary formats

Instead of using text-based formats like CSV, consider using binary formats like Parquet for reading and writing large DataFrames. Binary formats are more efficient in terms of storage and I/O performance.

```python
df = pd.read_parquet('data.parquet')
df.to_parquet('output.parquet')
```

By taking these practical steps, you can address common performance issues in Pandas and optimize your data analysis tasks for large datasets or complex operations. Keep these techniques in mind when working with Pandas to ensure that your code is efficient and performs well.

Summary

In this chapter, we discussed memory and computation efficiency in Pandas 2.0, focusing on the importance of optimizing performance for data analysis tasks. We also explored Dask as a solution for parallel and distributed computing with Pandas, and how to perform querying and filtering efficiently on the given customer_chrn_modelling database.

Next, we learned about vectorized operations, which are crucial for optimizing performance in Pandas. These operations allow you to perform element-wise calculations on entire arrays or DataFrames without the need for explicit loops, resulting in faster and more efficient code. We also discussed applying these operations to the given database.

We then explored the speed issues that can arise during data analysis, and how Cython and Numba can help resolve these issues. Both Cython and Numba can speed up performance by compiling Python code to native machine code or C. We walked through installing and configuring Cython and Numba, and showed how they can be used to optimize a simple example.

Finally, we discussed common performance issues that can arise when using Pandas and practical ways to resolve them. These issues include memory usage, non-vectorized operations, inefficient chaining of operations, slow groupby and aggregation operations, inefficient merging or joining operations, and slow reading and writing operations. By addressing these issues using techniques such as loading data in chunks, using appropriate data types, vectorizing operations, using the inplace parameter, optimizing the merge() method, using Dask, and choosing efficient file formats, you can significantly improve the efficiency and speed of your data analysis tasks with Pandas.

Chapter 7: Machine Learning with Pandas 2.0

Introduction to Machine Learning and Pandas

Machine learning (ML) is a subfield of artificial intelligence (AI) that focuses on the development of algorithms and models that allow computers to learn and make predictions or decisions based on input data without explicit programming. The learning process typically involves using a training dataset to "teach" the algorithm how to identify patterns, relationships, or structures within the data, and then using the learned knowledge to make predictions or decisions on new, unseen data.

Types of Machine Learning

There are three main types of machine learning:

Supervised learning: The algorithm is provided with labeled training data, which includes both input features and the corresponding output labels. The goal is to learn a mapping from input features to output labels so that the algorithm can predict the labels for new input data. Examples of supervised learning tasks include classification (e.g., identifying spam emails) and regression (e.g., predicting house prices).

Unsupervised learning: The algorithm is provided with unlabeled training data, which only contains input features without corresponding output labels. The goal is to discover patterns, structures, or relationships within the data without any prior knowledge about the outputs. Examples of unsupervised learning tasks include clustering (e.g., grouping customers based on their purchasing behavior) and dimensionality reduction (e.g., reducing the number of features while preserving the underlying structure of the data).

Reinforcement learning: The algorithm learns by interacting with an environment and receiving feedback in the form of rewards or penalties. The goal is to learn an optimal policy for making decisions that maximize the cumulative rewards over time. Examples of reinforcement learning tasks include game playing (e.g., training an agent to play chess) and robotics (e.g., training a robot to navigate a maze).

Pandas is a popular data manipulation and analysis library in Python that provides data structures and functions needed to work with structured data, such as tables or time series. While Pandas itself is not a machine learning library, it plays a significant role in the machine learning pipeline by providing essential tools for data preprocessing, feature engineering, and data analysis.

Role of Pandas in ML

The given below demonstrates how Pandas contributes to machine learning:

Data preprocessing: Cleaning and preparing the data is a crucial step in the machine learning pipeline. With Pandas, you can handle missing values, drop unnecessary columns, filter data based on conditions, and more.

Feature engineering: Creating new features or transforming existing features can help improve the performance of machine learning models. Pandas provides functions for performing operations such as scaling, normalization, binning, and encoding categorical variables.

Data analysis: Exploratory data analysis (EDA) is an essential step in understanding the data and identifying potential issues or patterns that can inform the choice of machine learning algorithms. Pandas provides tools for descriptive statistics, data visualization, and correlation analysis.

Data splitting: Before training a machine learning model, you need to split the data into training and testing sets. Pandas makes it easy to create random or stratified splits of the data.

Although Pandas is not a machine learning library itself, it plays a crucial role in the machine learning pipeline by providing the necessary tools for data manipulation and analysis. Once you've prepared the data using Pandas, you can use machine learning libraries like Scikit-learn, TensorFlow, or PyTorch to train and evaluate your models.

Data Preprocessing for Machine Learning

Data preprocessing is a crucial step in the machine learning pipeline, as it involves preparing and cleaning the data before feeding it to a machine learning model. We'll demonstrate how to perform data preprocessing using Pandas.

First, let us assume that you have loaded the dataset into a Pandas DataFrame:

```
import pandas as pd

url = 'https://raw.githubusercontent.com/kittenpub/database-repository/main/Customer_Churn_Modelling.csv'
```

```python
df = pd.read_csv(url)
```

Now, let us perform some common data preprocessing tasks:

Inspect the Data

Before preprocessing the data, it's important to get an understanding of the dataset, its features, and potential issues.

```python
# Display the first few rows of the DataFrame
print(df.head())


# Get summary statistics for numerical columns
print(df.describe())


# Check the data types of each column
print(df.dtypes)


# Check for missing values
print(df.isnull().sum())
```

Handle Missing Values

Depending on the nature of the missing values and the dataset, you can choose to fill them with a default value, interpolate, or drop the rows or columns with missing values.

```python
# Fill missing values with a default value (e.g., the mean of the column)
df['column_with_missing_values'] =
df['column_with_missing_values'].fillna(df['column_with_missing_values'].mean())


# Interpolate missing values
df.interpolate(method='linear', inplace=True)
```

```python
# Drop rows with missing values
df.dropna(inplace=True)

# Drop columns with a high percentage of missing values
df.drop('column_with_high_missing_percentage', axis=1, inplace=True)
```

Convert Categorical Data to Numerical

Machine learning algorithms typically require numerical input. Therefore, you need to convert categorical features to numerical values using techniques such as one-hot encoding or label encoding.

```python
# One-hot encoding
df = pd.get_dummies(df, columns=['categorical_column'])

# Label encoding
from sklearn.preprocessing import LabelEncoder

le = LabelEncoder()
df['categorical_column'] = le.fit_transform(df['categorical_column'])
```

Drop Unnecessary Columns

Some columns might not be useful for your analysis or might be highly correlated with other features. In such cases, you can drop those columns to simplify the dataset.

```python
df.drop(['unnecessary_column1', 'unnecessary_column2'], axis=1,
inplace=True)
```

Feature Scaling

Many machine learning algorithms are sensitive to the scale of the input features. To ensure that all features have the same scale, you can normalize or standardize the data.

```python
from sklearn.preprocessing import MinMaxScaler, StandardScaler
```

```python
# Normalization (Min-Max scaling)
min_max_scaler = MinMaxScaler()
df[['numerical_column1', 'numerical_column2']] =
min_max_scaler.fit_transform(df[['numerical_column1',
'numerical_column2']])

# Standardization (Z-score scaling)
standard_scaler = StandardScaler()
df[['numerical_column1', 'numerical_column2']] =
standard_scaler.fit_transform(df[['numerical_column1', 'numerical_column2']])
```

After completing these preprocessing steps, your dataset should be ready for use in a machine learning model. Remember that the specific preprocessing steps you choose to apply may vary depending on the dataset and the problem you're trying to solve.

Feature Engineering with Pandas 2.0

Feature engineering is the process of creating new features or transforming existing features in a dataset to improve the performance and predictive power of machine learning models. This process involves using domain knowledge, statistical techniques, and data analysis to extract more information from the raw data and provide meaningful input to machine learning algorithms. Effective feature engineering can lead to better model performance, reduced complexity, and improved interpretability.

Advantages of Feature Engineering

Improved model performance: Well-engineered features can capture important relationships and patterns in the data, making it easier for machine learning models to learn and generalize to new, unseen data.

Reduced complexity: By creating features that better represent the underlying structure of the data, you can often reduce the complexity of the machine learning model, which can lead to faster training times and better performance.

Interpretability: Creating meaningful features can make the resulting model more interpretable, which is important in certain applications where understanding the model's decision-making process is essential.

Handling categorical data: Machine learning models typically require numerical input. Feature engineering techniques such as one-hot encoding or label encoding can be used to convert categorical data into numerical values.

Handling missing values: Feature engineering can help in dealing with missing values, either by creating new features that capture the missingness pattern or by imputing missing values based on the relationships between features.

Dimensionality reduction: Techniques such as Principal Component Analysis (PCA) or other feature extraction methods can help reduce the number of features in a dataset, alleviating the curse of dimensionality and improving model performance.

Role of Pandas in Feature Engineering

Pandas is a powerful data manipulation library that provides various functions and tools that make it easier to perform feature engineering tasks. Some of the key functions and their use cases in feature engineering are:

Creating new features

Pandas allows you to create new features by applying mathematical operations, combining existing features, or using custom functions.

```python
# Create a new feature by applying a mathematical operation to an existing
feature
df['new_feature'] = df['existing_feature'] ** 2


# Create a new feature by combining two or more existing features
df['new_feature'] = df['feature1'] * df['feature2']


# Create a new feature using a custom function
def custom_function(x):
    return x * 2


df['new_feature'] = df['existing_feature'].apply(custom_function)
```

Encoding categorical variables

Pandas provides functions like get_dummies() and factorize() to convert categorical data into numerical values using one-hot encoding or label encoding.

```python
# One-hot encoding
df = pd.get_dummies(df, columns=['categorical_column'])

# Label encoding
df['categorical_column'], _ = pd.factorize(df['categorical_column'])
```

Binning and discretization

You can use Pandas functions like cut() and qcut() to discretize continuous variables into bins or categories, which can help capture non-linear relationships and reduce the impact of outliers.

```python
# Binning using equal-width bins
df['binned_column'] = pd.cut(df['continuous_column'], bins=10)

# Binning using equal-frequency bins
df['binned_column'] = pd.qcut(df['continuous_column'], q=10)
```

Handling missing values

Pandas provides functions like fillna(), interpolate(), and dropna() to handle missing values in a dataset through various techniques such as filling with a default value, interpolation, or dropping rows or columns with missing values.

```python
# Fill missing values with a default value (e.g., the mean of the column)
df['column_with_missing_values'] =
df['column_with_missing_values'].fillna(df['column_with_missing_values'].me
an())

Interpolate missing values
df.interpolate(method='linear', inplace=True)
```

Drop rows with missing values
df.dropna(inplace=True)

Drop columns with a high percentage of missing values
df.drop('column_with_high_missing_percentage', axis=1, inplace=True)

Feature scaling

Pandas can be used in conjunction with Scikit-learn's preprocessing tools to scale or normalize features, ensuring that all features have the same scale and preventing issues related to the scale of the input features.

```python
from sklearn.preprocessing import MinMaxScaler, StandardScaler

# Normalization (Min-Max scaling)
min_max_scaler = MinMaxScaler()
df[['numerical_column1', 'numerical_column2']] =
min_max_scaler.fit_transform(df[['numerical_column1',
'numerical_column2']])

# Standardization (Z-score scaling)
standard_scaler = StandardScaler()
df[['numerical_column1', 'numerical_column2']] =
standard_scaler.fit_transform(df[['numerical_column1', 'numerical_column2']])
```

Feature selection

Pandas can be used to compute correlation coefficients or other metrics that can help identify and remove highly correlated or irrelevant features, which can improve model performance and reduce overfitting.

```python
# Calculate the correlation matrix
corr_matrix = df.corr()

# Identify highly correlated features
```

```
highly_correlated_features = [column for column in corr_matrix.columns if
any(abs(corr_matrix[column]) > 0.9)]


# Drop highly correlated features
df.drop(highly_correlated_features, axis=1, inplace=True)
```

In summary, Pandas is an essential tool for feature engineering in machine learning, providing a wide range of functions and tools to create, transform, and manipulate features. By using Pandas effectively, you can extract more meaningful information from your data and improve the performance of your machine learning models.

Handling Imbalanced Data

Imbalanced data refers to a situation where the distribution of classes in the target variable is not equal or significantly skewed. In other words, one class has significantly more examples than the other classes. Imbalanced data can lead to biased machine learning models, as they tend to favor the majority class and perform poorly on the minority class.

To handle imbalanced data in the customer_chrn_modelling database, follow these steps:

Load Dataset and Explore Target Variable

```
import pandas as pd

url = 'https://raw.githubusercontent.com/kittenpub/database-
repository/main/Customer_Churn_Modelling.csv'
df = pd.read_csv(url)

# Assuming 'target_column' is the name of the target variable
print(df['target_column'].value_counts())
```

Identify Imbalanced Data

Check the distribution of classes in the target variable. If there's a significant difference between the number of examples in each class, the data is imbalanced.

There are several techniques to handle imbalanced data. Some common methods are:
- Resampling the data:
  - Upsampling the minority class: Increase the number of samples in the minority class by randomly replicating them.
  - Downsampling the majority class: Decrease the number of samples in the majority class by randomly removing them.
- Using synthetic data generation methods like Synthetic Minority Over-sampling Technique (SMOTE) or Adaptive Synthetic (ADASYN) to create new synthetic examples for the minority class.
- Using cost-sensitive learning, where you assign different misclassification costs to the majority and minority classes, effectively penalizing the model more for misclassifying the minority class.
- Ensemble methods like bagging and boosting can also help improve the performance on imbalanced datasets.

The given below demonstrates how to apply some of these techniques:

```python
# Import necessary libraries
from sklearn.utils import resample
from imblearn.over_sampling import SMOTE

# Separate the majority and minority classes
df_majority = df[df['target_column'] == 0]
df_minority = df[df['target_column'] == 1]

# Upsampling the minority class
df_minority_upsampled = resample(df_minority, replace=True,
n_samples=len(df_majority), random_state=42)
df_upsampled = pd.concat([df_majority, df_minority_upsampled])

# Downsampling the majority class
df_majority_downsampled = resample(df_majority, replace=False,
n_samples=len(df_minority), random_state=42)
df_downsampled = pd.concat([df_majority_downsampled, df_minority])
```

```
# Using SMOTE to generate synthetic samples for the minority class
X = df.drop('target_column', axis=1)
y = df['target_column']

smote = SMOTE(random_state=42)
X_smote, y_smote = smote.fit_resample(X, y)
```

After applying one or more of these techniques, you'll have a more balanced dataset that should help improve the performance of your machine learning models on the minority class. Note that it's essential to experiment with different methods and evaluate the performance of your model using appropriate metrics like precision, recall, F1-score, or area under the ROC curve, as accuracy can be misleading in the case of imbalanced data.

Feature Scaling and Normalization

Feature scaling and normalization are preprocessing techniques used to standardize the range of independent features or variables in a dataset. These techniques are essential because many machine learning algorithms are sensitive to the scale of input features. Different features might have different units and scales, and without proper scaling, an algorithm might give more importance to features with higher magnitudes, which can lead to biased or suboptimal model performance.

The two most common methods of feature scaling are normalization (Min-Max scaling) and standardization (Z-score scaling).

Normalization (Min-Max scaling)

Normalization scales the features in the range of [0, 1]. It is calculated using the following formula:

```
X_normalized = (X - X_min) / (X_max - X_min)
```

where X is the original feature value, X_min and X_max are the minimum and maximum values of the feature, respectively.

Standardization (Z-score scaling)

Standardization scales the features such that they have a mean of 0 and a standard deviation of 1. It is calculated using the following formula:

$$X\_standardized = (X - mean(X)) / std(X)$$

where X is the original feature value, mean(X) is the mean of the feature, and std(X) is the standard deviation of the feature.

Implementing Feature Scaling and Normalization

To implement feature scaling and normalization using Python, you can use the Scikit-learn library's preprocessing tools.

```python
import pandas as pd
from sklearn.preprocessing import MinMaxScaler, StandardScaler

# Load the dataset
url = 'https://raw.githubusercontent.com/kittenpub/database-
repository/main/Customer_Churn_Modelling.csv'
df = pd.read_csv(url)

# List the columns you want to scale
columns_to_scale = ['numerical_column1', 'numerical_column2']

# Normalization (Min-Max scaling)
min_max_scaler = MinMaxScaler()
df[columns_to_scale] = min_max_scaler.fit_transform(df[columns_to_scale])

# Standardization (Z-score scaling)
standard_scaler = StandardScaler()
df[columns_to_scale] = standard_scaler.fit_transform(df[columns_to_scale])
```

Remember to apply the same scaling method to both the training and test sets to ensure consistency in the data. However, make sure to fit the scaler only on the training data and use the same scaler to transform the test data. This avoids data leakage from the test set into the training set.

Train-Test Split and Cross-Validation

Train-Test Split and Cross-Validation are techniques used to evaluate the performance of machine learning models and ensure that they can generalize well to unseen data.

Train-Test Split

In train-test split, the dataset is divided into two separate sets: a training set and a test set. The training set is used to train the machine learning model, while the test set is used to evaluate the model's performance. The purpose of the train-test split is to provide an estimate of how well the model will perform on unseen data and to check for overfitting.

To implement train-test split using Scikit-learn, follow these steps:

```python
import pandas as pd
from sklearn.model_selection import train_test_split

# Load the dataset
url = 'https://raw.githubusercontent.com/kittenpub/database-
repository/main/Customer_Churn_Modelling.csv'
df = pd.read_csv(url)

# Separate the features and the target variable
X = df.drop('target_column', axis=1)
y = df['target_column']

# Perform train-test split
X_train, X_test, y_train, y_test = train_test_split(X, y, test_size=0.2,
random_state=42)
```

Cross-Validation

Cross-validation is a more robust technique to evaluate model performance. It involves partitioning the dataset into 'k' equal-sized folds, where 'k' is a user-defined parameter. In each iteration, the model is trained on 'k-1' folds and tested on the remaining fold. This process is repeated 'k' times, with each fold being used as the test set exactly once. The final performance metric is calculated as the average of the performance metrics from all 'k' iterations. Cross-validation helps to reduce the risk of overfitting and provides a more accurate estimate of model performance.

To implement cross-validation using Scikit-learn, follow these steps:

```python
import pandas as pd
from sklearn.model_selection import cross_val_score
from sklearn.linear_model import LogisticRegression

# Load the dataset
url = 'https://raw.githubusercontent.com/kittenpub/database-repository/main/Customer_Churn_Modelling.csv'
df = pd.read_csv(url)

# Separate the features and the target variable
X = df.drop('target_column', axis=1)
y = df['target_column']

# Create a model instance
model = LogisticRegression()

# Perform cross-validation with k=5 folds
scores = cross_val_score(model, X, y, cv=5)

# Calculate the average performance metric
mean_score = scores.mean()
```

Keep in mind that both techniques should be used together for a comprehensive evaluation of the model. The train-test split can be used for model selection and hyperparameter tuning, while cross-validation provides a more reliable estimate of the model's performance on unseen data.

Integration with Scikit-learn, TensorFlow, and PyTorch

Integrating Pandas with Scikit-learn, TensorFlow, and PyTorch allows for seamless data handling and preprocessing when working with these machine learning libraries.

Integrating Pandas with Scikit-learn

Scikit-learn is a widely used machine learning library in Python. The integration with Pandas is quite seamless as Scikit-learn can work directly with Pandas DataFrames. The given below is an example:

```python
import pandas as pd
from sklearn.model_selection import train_test_split
from sklearn.linear_model import LogisticRegression

# Load the dataset
url = 'https://raw.githubusercontent.com/kittenpub/database-
repository/main/Customer_Churn_Modelling.csv'
df = pd.read_csv(url)

# Separate the features and the target variable
X = df.drop('target_column', axis=1)
y = df['target_column']

# Perform train-test split
X_train, X_test, y_train, y_test = train_test_split(X, y, test_size=0.2,
random_state=42)

# Create a model instance and fit the model
model = LogisticRegression()
model.fit(X_train, y_train)

# Evaluate the model
```

```python
accuracy = model.score(X_test, y_test)
```

Integrating Pandas with TensorFlow

TensorFlow is an open-source library for machine learning and deep learning tasks. To use Pandas DataFrames with TensorFlow, you need to convert them into TensorFlow Datasets or NumPy arrays. The given below demonstrates an example using TensorFlow's Keras API:

```python
import pandas as pd
import tensorflow as tf
from sklearn.model_selection import train_test_split
from sklearn.preprocessing import StandardScaler

# Load the dataset
url = 'https://raw.githubusercontent.com/kittenpub/database-
repository/main/Customer_Churn_Modelling.csv'
df = pd.read_csv(url)

# Separate the features and the target variable
X = df.drop('target_column', axis=1).values
y = df['target_column'].values

# Perform train-test split
X_train, X_test, y_train, y_test = train_test_split(X, y, test_size=0.2,
random_state=42)

# Scale the features
scaler = StandardScaler()
X_train = scaler.fit_transform(X_train)
X_test = scaler.transform(X_test)

# Build and train a model using Keras
```

```python
model = tf.keras.Sequential([
    tf.keras.layers.Dense(16, activation='relu', input_shape=(X_train.shape[1],)),
    tf.keras.layers.Dense(1, activation='sigmoid')
])

model.compile(optimizer='adam', loss='binary_crossentropy',
metrics=['accuracy'])
model.fit(X_train, y_train, epochs=10, batch_size=32, validation_split=0.1)

# Evaluate the model
_, accuracy = model.evaluate(X_test, y_test)
```

Integrating Pandas with PyTorch

PyTorch is an open-source machine learning library for Python, based on Torch. To use Pandas DataFrames with PyTorch, you need to convert them into PyTorch Tensors. The given below is an example:

```python
import pandas as pd
import torch
from torch.utils.data import DataLoader, TensorDataset
from sklearn.model_selection import train_test_split
from sklearn.preprocessing import StandardScaler

# Load the dataset
url = 'https://raw.githubusercontent.com/kittenpub/database-
repository/main/Customer_Churn_Modelling.csv'
df = pd.read_csv(url)

# Separate the features and the target variable
X = df.drop('target_column', axis=1).values
y = df['target_column'].values
```

```python
# Perform train-test split
X_train, X
# Scale the features
scaler = StandardScaler()
X_train = scaler.fit_transform(X_train)
X_test = scaler.transform(X_test)

Convert the NumPy arrays into PyTorch Tensors and create DataLoaders
train_dataset = TensorDataset(torch.Tensor(X_train), torch.Tensor(y_train))
test_dataset = TensorDataset(torch.Tensor(X_test), torch.Tensor(y_test))

train_loader = DataLoader(train_dataset, batch_size=32, shuffle=True)
test_loader = DataLoader(test_dataset, batch_size=32)

Build and train a model using PyTorch
model = torch.nn.Sequential(
torch.nn.Linear(X_train.shape[1], 16),
torch.nn.ReLU(),
torch.nn.Linear(16, 1),
torch.nn.Sigmoid()
)

criterion = torch.nn.BCELoss()
optimizer = torch.optim.Adam(model.parameters())

for epoch in range(10):
for X_batch, y_batch in train_loader:
y_pred = model(X_batch)
loss = criterion(y_pred, y_batch.view(-1, 1))
optimizer.zero_grad()
loss.backward()
optimizer.step()
```

```python
Evaluate the model
with torch.no_grad():
correct = 0
total = 0
for X_batch, y_batch in test_loader:
y_pred = model(X_batch)
predicted = (y_pred > 0.5).float()
total += y_batch.size(0)
correct += (predicted == y_batch.view(-1, 1)).sum().item()

accuracy = correct / total
```

In summary, integrating Pandas with Scikit-learn, TensorFlow, and PyTorch allows for a streamlined workflow in machine learning tasks, as Pandas provides powerful data manipulation tools while these libraries provide advanced machine learning and deep learning functionalities.

Summary

We covered topics related to machine learning with pandas, including data preprocessing, feature engineering, handling imbalanced data, and feature scaling. We also covered the techniques of train-test split and cross-validation, along with examples of how to implement them using Python and Scikit-learn.

Furthermore, we discussed the integration of pandas with various machine learning libraries, including Scikit-learn, TensorFlow, and PyTorch, which allows for seamless data handling and preprocessing in machine learning tasks. We provided examples of how to integrate pandas with each of these libraries and implement various machine learning tasks.

One of the essential aspects of machine learning is feature engineering, which involves selecting, transforming, and engineering features to improve model performance. We discussed various feature engineering techniques, such as one-hot encoding, feature scaling, and dimensionality reduction, along with their advantages and benefits for machine learning.

In addition, we talked about handling imbalanced data, which is a common issue in machine learning, especially in classification tasks. We provided step-by-step instructions on how to identify and resolve imbalanced data using Python and Scikit-learn.

Finally, we covered feature scaling and normalization, which are preprocessing techniques used to standardize the range of independent features or variables in a dataset. We discussed the importance of feature scaling in machine learning and provided examples of how to implement feature scaling and normalization using Scikit-learn.

Overall, the chapter covered various essential topics related to machine learning with pandas, including data preprocessing, feature engineering, handling imbalanced data, and model evaluation techniques. We also demonstrated the integration of pandas with various machine learning libraries, which is essential for a streamlined workflow in machine learning tasks.

CHAPTER 8: TEXT DATA AND NATURAL LANGUAGE PROCESSING

Text Data Cleaning and Preprocessing

Working with text data is an important part of data analysis, as many datasets contain textual information that needs to be processed and analyzed. Text data can come in many different forms, such as customer reviews, social media posts, or even medical records.

In data analysis, working with text data involves a number of tasks, including cleaning and preprocessing the data, extracting relevant information, and analyzing patterns and trends within the data. To accomplish these tasks, data analysts often use specialized tools and techniques that are designed specifically for working with text data.

One of the most powerful tools for working with text data in Python is the Pandas library. Pandas provides a number of string methods that are specifically designed for working with text data in a DataFrame or Series. These string methods make it easy to manipulate and analyze text data, allowing data analysts to quickly and easily extract relevant information and gain insights from their data.

The following are some of the most commonly used string methods in Pandas for working with text data:

str.contains()

The str.contains() method is used to check whether a string contains a specified substring and returns a Boolean value. The given below is an example of how to use it:

```python
import pandas as pd

# create a sample DataFrame
df = pd.DataFrame({'text': ['hello world', 'python is awesome', 'data science is cool']})

# check whether the 'text' column contains the substring 'cool'
df['contains_cool'] = df['text'].str.contains('cool')

# display the updated DataFrame
print(df)
```

Output:

```
            text  contains_cool
0        hello world        False
1   python is awesome        False
2  data science is cool       True
```

As you can see, the str.contains() method has added a new column to the DataFrame that indicates whether each row in the 'text' column contains the substring 'cool' or not.

str.startswith()

The str.startswith() method is used to check whether a string starts with a specified substring and returns a Boolean value. The given below is an example of how to use it:

```python
import pandas as pd

# create a sample DataFrame
df = pd.DataFrame({'text': ['hello world', 'python is awesome', 'data science is cool']})

# check whether the 'text' column starts with the substring 'hello'
df['starts_with_hello'] = df['text'].str.startswith('hello')

# display the updated DataFrame
print(df)
```

Output:

```
            text  starts_with_hello
0        hello world         True
1   python is awesome        False
2  data science is cool       False
```

As you can see, the str.startswith() method has added a new column to the DataFrame that indicates whether each row in the 'text' column starts with the substring 'hello' or not.

str.endswith()

The str.endswith() method is used to check whether a string ends with a specified substring and returns a Boolean value. The given below is an example of how to use it:

```python
import pandas as pd

# create a sample DataFrame
df = pd.DataFrame({'text': ['hello world', 'python is awesome', 'data science is cool']})

# check whether the 'text' column ends with the substring 'world'
df['ends_with_world'] = df['text'].str.endswith('world')

# display the updated DataFrame
print(df)
```

Output:

```
              text  ends_with_world
0      hello world             True
1  python is awesome            False
2  data science is cool         False
```

As you can see, the str.endswith() method has added a new column to the DataFrame that indicates whether each row in the 'text' column ends with the substring 'world' or not.

str.split()

The str.split() method is used to split a string into a list of substrings based on a specified separator. The given below demonstrates an example of how to use it:

```python
import pandas as pd
```

```python
# create a sample DataFrame
df = pd.DataFrame({'text': ['hello,world', 'python,is,awesome',
'data,science,is,cool']})

# split the 'text' column into a list of substrings using a comma as the separator
df['text_list'] = df['text'].str.split(',')

# display the updated DataFrame
print(df)
```

Output:

	text	text_list
0	hello,world	[hello, world]
1	python,is,awesome	[python,

As we can see from the previous example, the str.split() method has split the 'text' column into a list of substrings based on the comma separator. The resulting list is then stored in a new column called 'text_list'.

str.strip()

The str.strip() method is used to remove leading and trailing whitespace from a string. The given below demonstrates an example of how to use it:

```python
import pandas as pd

# create a sample DataFrame
df = pd.DataFrame({'text': ['  hello world  ', ' python is awesome ', ' data
science is cool  ']})

# remove leading and trailing whitespace from the 'text' column
df['text_stripped'] = df['text'].str.strip()
```

```python
# display the updated DataFrame
print(df)
```

Output:

```
            text          text_stripped
0      hello world           hello world
1    python is awesome     python is awesome
2  data science is cool   data science is cool
```

As we can see from the output, the str.strip() method has removed leading and trailing whitespace from the 'text' column and stored the cleaned text in a new column called 'text_stripped'.

str.replace()

The str.replace() method is used to replace a specified substring with another substring. The given below demonstrates an example of how to use it:

```python
import pandas as pd

# create a sample DataFrame
df = pd.DataFrame({'text': ['hello world', 'python is awesome', 'data science is cool']})

# replace 'world' with 'universe' in the 'text' column
df['text_replaced'] = df['text'].str.replace('world', 'universe')

# display the updated DataFrame
print(df)
```

Output:

```
          text        text_replaced
0      hello world       hello universe
1    python is awesome    python is awesome
2 data science is cool  data science is cool
```

As we can see from the output, the str.replace() method has replaced the substring 'world' with 'universe' in the 'text' column and stored the updated text in a new column called 'text_replaced'.

str.lower()

The str.lower() method is used to convert a string to lowercase. The given below walkthrough an example of how to use it:

```python
import pandas as pd

# create a sample DataFrame
df = pd.DataFrame({'text': ['Hello World', 'Python is Awesome', 'Data Science is Cool']})

# convert the 'text' column to lowercase
df['text_lower'] = df['text'].str.lower()

# display the updated DataFrame
print(df)
```

Output:

```
          text            text_lower
0      Hello World         hello world
1    Python is Awesome      python is awesome
2 Data Science is Cool    data science is cool
```

As we can see from the output, the str.lower() method has converted the 'text' column to lowercase and stored the lowercase text in a new column called 'text_lower'.

str.upper()

The str.upper() method is used to convert a string to uppercase. The given below walkthrough an example of how to use it:

```python
import pandas as pd

# create a sample DataFrame
df = pd.DataFrame({'text': ['Hello World', 'Python is Awesome', 'Data Science is Cool']})

# convert the 'text' column to uppercase
df['text_upper'] = df['text'].str.upper()

# display the updated DataFrame
print(df)
```

Output:

	text	text_upper
0	Hello World	HELLO WORLD

As we can see from the output, the str.upper() method has converted the 'text' column to uppercase and stored the uppercase text in a new column called 'text_upper'.

str.len()

The str.len() method is used to return the length of a string. The given below walkthrough an example of how to use it:

```python
import pandas as pd

# create a sample DataFrame
df = pd.DataFrame({'text': ['hello world', 'python is awesome', 'data science is
```

```python
cool']})

# get the length of each string in the 'text' column
df['text_length'] = df['text'].str.len()

# display the updated DataFrame
print(df)
```

Output:

```
              text  text_length
0       hello world           11
1  python is awesome           18
2  data science is cool        22
```

As we can see from the output, the str.len() method has returned the length of each string in the 'text' column and stored the result in a new column called 'text_length'.

str.isnumeric()

The str.isnumeric() method is used to check whether a string consists only of numeric characters and returns a Boolean value. The given below walkthrough an example of how to use it:

```python
import pandas as pd

# create a sample DataFrame
df = pd.DataFrame({'text': ['12345', '1a2b3c4d5e', '123 45']})

# check whether each string in the 'text' column consists only of numeric
characters
df['is_numeric'] = df['text'].str.isnumeric()

# display the updated DataFrame
```

```python
print(df)
```

Output:

```
        text  is_numeric
0       12345        True
1  1a2b3c4d5e       False
2      123 45       False
```

As we can see from the output, the str.isnumeric() method has checked whether each string in the 'text' column consists only of numeric characters and stored the result in a new column called 'is_numeric'.

str.capitalize()

The str.capitalize() method is used to capitalize the first letter of a string. The given below walkthrough an example of how to use it:

```python
import pandas as pd

# create a sample DataFrame
df = pd.DataFrame({'text': ['hello world', 'python is awesome', 'data science is cool']})

# capitalize the first letter of each string in the 'text' column
df['text_capitalized'] = df['text'].str.capitalize()

# display the updated DataFrame
print(df)
```

Output:

```
          text  text_capitalized
0  hello world       Hello world
```

```
1     python is awesome   Python is awesome
2 data science is cool  Data science is cool
```

As we can see from the output, the str.capitalize() method has capitalized the first letter of each string in the 'text' column and stored the result in a new column called 'text_capitalized'.

str.title()

The str.title() method is used to capitalize the first letter of each word in a string. The given below walkthrough an example of how to use it:

```python
import pandas as pd

# create a sample DataFrame
df = pd.DataFrame({'text': ['hello world', 'python is awesome', 'data science is cool']})

# capitalize the first letter of each word in each string in the 'text' column
df['text_title'] = df['text'].str.title()

# display the updated DataFrame
print(df)
```

Output:

```
            text        text_title
0        hello world       Hello World
1     python is awesome     Python Is Awesome
2 data science is cool  Data Science Is Cool
```

As we can see from the output, the str.title() method has capitalized the first letter of each word in each string in the 'text' column and stored the result in a new column called 'text_title'.

These are just some of the many string methods available in Pandas for working with text data. By using these methods, data analysts can quickly and easily clean, preprocess, and analyze text data in a DataFrame or Series.

Extracting and Transforming Text Features

Following is an example of how to perform text data cleaning and preprocessing on the training data provided in the CSV file located at:
https://github.com/kittenpub/database-repository/blob/main/pandas_training_testdata.csv

```python
import pandas as pd

# load the training data into a DataFrame
df = pd.read_csv('https://raw.githubusercontent.com/kittenpub/database-repository/main/pandas_training_testdata.csv')

# check the shape of the DataFrame
print("Shape of the DataFrame:", df.shape)

# check the first few rows of the DataFrame
print("First few rows of the DataFrame:")
print(df.head())

# check for missing values
print("Number of missing values in each column:")
print(df.isnull().sum())

# remove rows with missing values
df.dropna(inplace=True)

# convert text to lowercase
df['text'] = df['text'].str.lower()
```

```python
# remove punctuation
import string
df['text'] = df['text'].apply(lambda x: x.translate(str.maketrans('', '',
string.punctuation)))

# remove numbers
df['text'] = df['text'].str.replace('\d+', '')

# remove stopwords
import nltk
nltk.download('stopwords')
from nltk.corpus import stopwords
stop_words = stopwords.words('english')
df['text'] = df['text'].apply(lambda x: ' '.join([word for word in x.split() if word
not in stop_words]))

# remove short words
df['text'] = df['text'].apply(lambda x: ' '.join([word for word in x.split() if
len(word) > 3]))

# lemmatize the text
nltk.download('wordnet')
from nltk.stem import WordNetLemmatizer
lemmatizer = WordNetLemmatizer()
df['text'] = df['text'].apply(lambda x: ' '.join([lemmatizer.lemmatize(word) for
word in x.split()]))

# check the first few rows of the cleaned DataFrame
print("First few rows of the cleaned DataFrame:")
print(df.head())

# save the cleaned DataFrame to a new CSV file
```

```
df.to_csv('cleaned_data.csv', index=False)
```

In this example, we first load the training data into a DataFrame using the read_csv() method. We then check the shape of the DataFrame and the first few rows to get an idea of the data.

Next, we check for missing values using the isnull() method and remove rows with missing values using the dropna() method.

We then perform various text data cleaning and preprocessing steps on the 'text' column of the DataFrame. Specifically, we convert the text to lowercase using the lower() method, remove punctuation using the translate() method, remove numbers using the str.replace() method, remove stopwords using the NLTK library's stopwords corpus, remove short words using a lambda function, and lemmatize the text using the NLTK library's WordNetLemmatizer.

Finally, we check the first few rows of the cleaned DataFrame and save the cleaned DataFrame to a new CSV file using the to_csv() method.

Sentiment Analysis with Text Data

Sentiment analysis is a process of analyzing and extracting the sentiment or emotion expressed in a piece of text. It is also known as opinion mining or emotion AI. The goal of sentiment analysis is to determine whether the sentiment expressed in a piece of text is positive, negative, or neutral.

In essence, sentiment analysis involves two main steps: text preprocessing and sentiment classification. Text preprocessing involves cleaning and transforming the raw text data to make it suitable for analysis. This may involve tasks such as tokenization, removing stop words, stemming or lemmatization, and more. Sentiment classification involves applying a machine learning or natural language processing model to the preprocessed text to classify it as positive, negative, or neutral.

Pandas can be used to load and preprocess the text data, but more advanced machine learning or natural language processing libraries such as Scikit-learn, NLTK, or spaCy would be required for sentiment classification.

Below is a step-by-step explanation of the practical demonstration of sentiment analysis using Pandas and Scikit-learn:

Load Data into Pandas DataFrame

```
import pandas as pd

# load the data into a DataFrame
df = pd.read_csv('https://raw.githubusercontent.com/kittenpub/database-
repository/main/sentiment_analysis_data.csv')
```

In this step, we use the read_csv() method from Pandas to read the CSV file containing the text data into a DataFrame.

Preprocess the Text Data

```
import string
import nltk
from nltk.corpus import stopwords
from nltk.stem import WordNetLemmatizer

# convert the text to lowercase
df['text'] = df['text'].str.lower()

# remove punctuation
df['text'] = df['text'].apply(lambda x: x.translate(str.maketrans('', '',
string.punctuation)))

# remove stop words
stop_words = stopwords.words('english')
df['text'] = df['text'].apply(lambda x: ' '.join([word for word in x.split() if word
not in stop_words]))

# lemmatize the text
lemmatizer = WordNetLemmatizer()
```

```python
df['text'] = df['text'].apply(lambda x: ' '.join([lemmatizer.lemmatize(word) for word in x.split()]))
```

In this step, we perform several preprocessing steps on the 'text' column of the DataFrame. We first convert the text to lowercase using the lower() method. We then remove punctuation using the translate() method from the string module. We remove stop words using the stopwords corpus from the nltk library. Finally, we lemmatize the text using the WordNetLemmatizer from the nltk library.

Convert Text Data to Sparse Matrix of Word Counts

```python
from sklearn.feature_extraction.text import CountVectorizer

# create a CountVectorizer object
vectorizer = CountVectorizer()

# fit the vectorizer on the text data to learn the vocabulary
vectorizer.fit(df['text'])

# transform the text data into a sparse matrix of word counts
X = vectorizer.transform(df['text'])
```

In this step, we create a CountVectorizer object from Scikit-learn to convert the text data into a sparse matrix of word counts. We fit the vectorizer on the 'text' column of the DataFrame to learn the vocabulary, and then transform the 'text' column into a sparse matrix of word counts using the transform() method.

Split Data into Training and Testing Sets

```python
from sklearn.model_selection import train_test_split

# split the data into training and testing sets
X_train, X_test, y_train, y_test = train_test_split(X, df['sentiment'],
```

```
test_size=0.2, random_state=42)
```

In this step, we use the train_test_split() method from Scikit-learn to split the data into training and testing sets. We use 80% of the data for training and 20% for testing.

Train Machine Learning Model on Training Data

```
from sklearn.naive_bayes import MultinomialNB

# train a Multinomial Naive Bayes classifier on the training data
clf = MultinomialNB()
clf.fit(X_train, y_train)
```

In this step, we train a Multinomial Naive Bayes classifier on the training data using the MultinomialNB() class from Scikit-learn.

Make Predictions and Evaluate Model's Accuracy

```
from sklearn.metrics import accuracy_score

# make predictions on the testing data
y_pred = clf.predict(X_test)

# evaluate the accuracy of the model
accuracy = accuracy_score(y_test, y_pred)
print("Accuracy:", accuracy)
```

In this step, we use the predict() method from the trained classifier to make predictions on the testing data. We then evaluate the accuracy of the model using the accuracy_score() method from Scikit-learn. The accuracy score represents the percentage of correctly classified instances in the testing data.

<u>Visualize the Results</u>

```python
import matplotlib.pyplot as plt
import seaborn as sns

# create a confusion matrix
from sklearn.metrics import confusion_matrix
conf_mat = confusion_matrix(y_test, y_pred)

# plot the confusion matrix
sns.heatmap(conf_mat, annot=True, cmap='Blues', fmt='d',
        xticklabels=['negative', 'neutral', 'positive'],
        yticklabels=['negative', 'neutral', 'positive'])
plt.xlabel('Predicted label')
plt.ylabel('True label')
plt.title('Confusion matrix')
plt.show()
```

In this step, we use the confusion_matrix() method from Scikit-learn to create a confusion matrix of the predicted and true labels. We then use the heatmap() method from the seaborn library to visualize the confusion matrix as a heatmap.

Overall, the above steps demonstrate how to perform sentiment analysis using Pandas and Scikit-learn. The steps involve loading the data into a Pandas DataFrame, preprocessing the text data, converting the text data into a sparse matrix of word counts, splitting the data into training and testing sets, training a machine learning model on the training data, making predictions on the testing data, and evaluating the model's accuracy. The results can also be visualized using a confusion matrix.

Topic Modeling

Topic modeling and text clustering are two common techniques used in natural language processing to analyze and organize large collections of text data. Both techniques are used to extract useful information from text data and make it easier to understand and analyze.

Topic modeling is a technique used to identify topics or themes present in a large collection of text documents. The goal of topic modeling is to automatically identify the underlying topics that are discussed in a set of documents without any prior knowledge of the topics. Topic modeling algorithms use statistical methods to group words that frequently co-occur in documents and identify topics that can be described by these word groups.

One of the most popular algorithms used for topic modeling is Latent Dirichlet Allocation (LDA). LDA is a generative probabilistic model that assumes each document is a mixture of topics and each topic is a distribution over words. LDA estimates the topic distribution and word distribution for each document and topic by iteratively estimating the probability distribution of the words.

Process of Implementing Topic Modeling

The process of performing topic modeling using Latent Dirichlet Allocation (LDA) involves several steps:

Load and Preprocess Text Data

The first step is to load the text data into a format suitable for analysis. This may involve cleaning the text data, removing stop words, stemming or lemmatization, and more. The preprocessed text data is typically represented as a collection of documents, where each document is a list of words.

Create a Document-term Matrix

The next step is to create a document-term matrix, which represents the frequency of each term (word) in each document. This matrix is typically represented as a sparse matrix to save memory.

Choose Number of Topics

The number of topics in the text data must be determined before applying the LDA algorithm. This is often done using domain knowledge or trial and error.

Train the LDA Model

The LDA model is trained on the document-term matrix using an iterative algorithm. The model estimates the topic distribution and word distribution for each document and topic by iteratively estimating the probability distribution of the words. The output of the LDA model is a set of topics, each represented as a distribution over words.

Evaluate the LDA Model

The quality of the LDA model can be evaluated using metrics such as coherence or perplexity. Coherence measures the degree of semantic similarity between the words in a topic, while perplexity measures how well the model predicts unseen data.

Interpret the Topics

The final step is to interpret the topics and assign human-readable labels to them based on the most frequent words in each topic. This step may involve manual inspection of the topics and some domain knowledge.

LDA is a powerful technique for identifying topics or themes present in a large collection of text documents and can be used in a variety of applications such as information retrieval, data mining, and content analysis.

Performing Topic Modeling using Latent Dirichlet Allocation (LDA)

Below is a practical demonstration of how to perform topic modeling using Latent Dirichlet Allocation (LDA) in Pandas 2.0:

Load and Preprocess Text Data

```python
import pandas as pd
import string
from nltk.corpus import stopwords
from nltk.stem import WordNetLemmatizer

# load the data into a DataFrame
df = pd.read_csv('https://raw.githubusercontent.com/kittenpub/database-
repository/main/topic_modeling_data.csv')

# convert the text to lowercase
df['text'] = df['text'].str.lower()

# remove punctuation
df['text'] = df['text'].apply(lambda x: x.translate(str.maketrans('', '',
string.punctuation)))
```

```python
# remove stop words
stop_words = stopwords.words('english')
df['text'] = df['text'].apply(lambda x: ' '.join([word for word in x.split() if word
not in stop_words]))

# lemmatize the text
lemmatizer = WordNetLemmatizer()
df['text'] = df['text'].apply(lambda x: ' '.join([lemmatizer.lemmatize(word) for
word in x.split()]))
```

In this step, we load the data into a DataFrame using the read_csv() method from Pandas.
We then perform several preprocessing steps on the 'text' column of the DataFrame. We
convert the text to lowercase using the lower() method. We then remove punctuation using
the translate() method from the string module. We remove stop words using the stopwords
corpus from the nltk library. Finally, we lemmatize the text using the WordNetLemmatizer
from the nltk library.

Create a Document-term Matrix

```python
from sklearn.feature_extraction.text import CountVectorizer

# create a CountVectorizer object
vectorizer = CountVectorizer()

# fit the vectorizer on the text data to learn the vocabulary
vectorizer.fit(df['text'])

# transform the text data into a document-term matrix
doc_term_matrix = vectorizer.transform(df['text'])
```

In this step, we create a CountVectorizer object from Scikit-learn to convert the text data
into a document-term matrix. We fit the vectorizer on the 'text' column of the DataFrame
to learn the vocabulary, and then transform the 'text' column into a document-term matrix

using the transform() method.

Train the LDA Model

```
# choose the number of topics
num_topics = 5

# create an LDA model
lda_model = LatentDirichletAllocation(n_components=num_topics,
random_state=42)

# fit the LDA model on the document-term matrix
lda_model.fit(doc_term_matrix)
```

In this step, we choose the number of topics in the text data (in this case, 5) and create an LDA model using the LatentDirichletAllocation class from Scikit-learn. We fit the LDA model on the document-term matrix using the fit() method.

Interpret the Topics

```
# print the top 10 words in each topic
feature_names = vectorizer.get_feature_names()
for topic_idx, topic in enumerate(lda_model.components_):
    print(f"Topic {topic_idx}:")
    print(" ".join([feature_names[i] for i in topic.argsort()[:-10 - 1:-1]]))
    print()
```

In this step, we interpret the topics by printing the top 10 words in each topic. We use the get_feature_names() method from the CountVectorizerobject to get the list of feature names (i.e., words), and then use theargsort() method to sort the topics by the frequency of the words. Finally, we print the top 10 words in each topic.

<u>Evaluate the LDA Model</u>

```
# print the perplexity score
print("Perplexity:", lda_model.perplexity(doc_term_matrix))
```

In this step, we evaluate the LDA model by printing the perplexity score using the perplexity() method from the trained LDA model. The perplexity score measures how well the model predicts unseen data, and a lower score indicates a better model.

Overall, the above steps demonstrate how to perform topic modeling using LDA in Pandas 2.0. The steps involve loading and preprocessing the text data, creating a document-term matrix, training the LDA model, interpreting the topics, and evaluating the model.

Text Clustering

Text clustering, on the other hand, is a technique used to group similar documents together based on their content. The goal of text clustering is to automatically identify groups of documents that share common themes or topics. Text clustering algorithms use unsupervised machine learning techniques to group similar documents based on their content.

One of the most popular algorithms used for text clustering is K-means clustering. K-means clustering is an iterative algorithm that groups documents into K clusters based on the similarity of their content. The algorithm iteratively assigns each document to the closest cluster based on a distance metric such as Euclidean distance or cosine similarity.

Both topic modeling and text clustering can be useful in a variety of applications such as information retrieval, data mining, and content analysis. They can be used to identify patterns and trends in large collections of text data, classify documents based on their content, and summarize the information contained in a large set of documents.

<u>Text Clustering Procedure</u>

The process of performing text clustering using K-means involves several steps:

Load and preprocess the text data
The first step is to load the text data into a format suitable for analysis. This may involve cleaning the text data, removing stop words, stemming or lemmatization, and more. The preprocessed text data is typically represented as a collection of documents, where each

document is a list of words.

Create a document-term matrix

The next step is to create a document-term matrix, which represents the frequency of each term (word) in each document. This matrix is typically represented as a sparse matrix to save memory.

Choose the number of clusters

The number of clusters in the text data must be determined before applying the K-means algorithm. This is often done using domain knowledge or trial and error.

Train the K-means model

The K-means model is trained on the document-term matrix using an iterative algorithm. The algorithm iteratively assigns each document to the closest cluster based on a distance metric such as Euclidean distance or cosine similarity. The output of the K-means model is a set of clusters, each represented as a centroid vector.

Evaluate the K-means model

The quality of the K-means model can be evaluated using metrics such as silhouette score or inertia. Silhouette score measures how well the instances are clustered together, while inertia measures how tightly the instances are clustered.

Interpret the clusters

The final step is to interpret the clusters and assign human-readable labels to them based on the most frequent words in each cluster. This step may involve manual inspection of the clusters and some domain knowledge.

K-means is a powerful technique for grouping similar documents together based on their content and can be used in a variety of applications such as information retrieval, data mining, and content analysis.

Performing Text Clustering using K-means

here's a practical demonstration of how to perform text clustering using K-means in Pandas 2.0:

Load and Preprocess Text Data

```python
import pandas as pd
import string
from nltk.corpus import stopwords
from nltk.stem import WordNetLemmatizer

# load the data into a DataFrame
df = pd.read_csv('https://raw.githubusercontent.com/kittenpub/database-
repository/main/text_clustering_data.csv')

# convert the text to lowercase
df['text'] = df['text'].str.lower()

# remove punctuation
df['text'] = df['text'].apply(lambda x: x.translate(str.maketrans('', '',
string.punctuation)))

# remove stop words
stop_words = stopwords.words('english')
df['text'] = df['text'].apply(lambda x: ' '.join([word for word in x.split() if word
not in stop_words]))

# lemmatize the text
lemmatizer = WordNetLemmatizer()
df['text'] = df['text'].apply(lambda x: ' '.join([lemmatizer.lemmatize(word) for
word in x.split()]))
```

In this step, we load the data into a DataFrame using the read_csv() method from Pandas. We then perform several preprocessing steps on the 'text' column of the DataFrame. We convert the text to lowercase using the lower() method. We then remove punctuation using the translate() method from the string module. We remove stop words using the stopwords corpus from the nltk library. Finally, we lemmatize the text using the WordNetLemmatizer from the nltk library.

Create Document-term Matrix

```python
from sklearn.feature_extraction.text import CountVectorizer

# create a CountVectorizer object
vectorizer = CountVectorizer()

# fit the vectorizer on the text data to learn the vocabulary
vectorizer.fit(df['text'])

# transform the text data into a document-term matrix
doc_term_matrix = vectorizer.transform(df['text'])
```

In this step, we create a CountVectorizer object from Scikit-learn to convert the text data into a document-term matrix. We fit the vectorizer on the 'text' column of the DataFrame to learn the vocabulary, and then transform the 'text' column into a document-term matrix using the transform() method.

Train the K-means Model

```python
from sklearn.cluster import KMeans

# choose the number of clusters
num_clusters = 3

# create a K-means model
kmeans_model = KMeans(n_clusters=num_clusters, random_state=42)

# fit the K-means model on the document-term matrix
kmeans_model.fit(doc_term_matrix)
```

In this step, we choose the number of clusters in the text data (in this case, 3) and create a K-means model using the KMeans class from Scikit-learn. We fit the K-means model on the document-term matrix using the fit() method.

Interpret the Clusters

```
# get the top 10 terms in each cluster
feature_names = vectorizer.get_feature_names()
for i in range(num_clusters):
    print(f"Cluster {i}:")
    top_terms = [feature_names[idx] for idx in
kmeans_model.cluster_centers_[i].argsort()[:-10 - 1:-1]]
    print(", ".join(top_terms))
    print()
```

In this step, we interpret the clusters by printing the top 10 terms in each cluster. We use the get_feature_names() method from the CountVectorizer object to get the list of feature names (i.e., words), and then use the argsort() method and cluster_centers_ attribute from the trained K-means model to sort the terms by their frequencies in each cluster. Finally, we print the top 10 terms in each cluster.

Evaluate the K-means Model

```
from sklearn.metrics import silhouette_score

# calculate the silhouette score
silhouette_score = silhouette_score(doc_term_matrix, kmeans_model.labels_)
print("Silhouette Score:", silhouette_score)
```

In this step, we evaluate the K-means model by calculating the silhouette score using the silhouette_score() method from Scikit-learn. The silhouette score measures how well the instances are clustered together, and a higher score indicates a better model.

Overall, the above steps demonstrate how to perform text clustering using K-means in Pandas 2.0. The steps involve loading and preprocessing the text data, creating a document-term matrix, training the K-means model, interpreting the clusters, and evaluating the model. K-means is a powerful technique for grouping similar documents together based on their content and can be used in a variety of applications such as information retrieval, data mining, and content analysis.

Summary

In this chapter, we covered several topics related to text data analysis using Pandas 2.0. We started by discussing text data cleaning and preprocessing, which involves several steps such as converting the text to lowercase, removing punctuation, removing stop words, and lemmatization. We demonstrated how to perform text data cleaning and preprocessing on a sample dataset using Pandas 2.0.

We then moved on to sentiment analysis, which involves determining the sentiment (positive, negative, or neutral) of a given text. We explained the process of performing sentiment analysis using Pandas 2.0, which involves loading and preprocessing the text data, training a machine learning model, and evaluating the model's performance. We demonstrated how to perform sentiment analysis on a sample dataset using Pandas 2.0 and Scikit-learn.

Next, we discussed topic modeling and text clustering, which are techniques for grouping similar documents together based on their content. We explained the process of performing topic modeling using Latent Dirichlet Allocation (LDA) and text clustering using K-means, which involve several steps such as creating a document-term matrix, training a machine learning model, and interpreting the results. We demonstrated how to perform topic modeling and text clustering on a sample dataset using Pandas 2.0 and Scikit-learn.

CHAPTER 9:
GEOSPATIAL DATA ANALYSIS

Introduction to Geospatial Data and Pandas 2.0

Geospatial data, also known as spatial data or geographic data, refers to any data that has a location component attached to it. Geospatial data can be in the form of points (e.g., latitude and longitude coordinates), lines (e.g., roads and rivers), or polygons (e.g., land parcels and administrative boundaries). Geospatial data can come from various sources such as GPS devices, satellite imagery, and maps.

Working with geospatial data in data analysis involves several steps. The first step is to acquire the geospatial data and load it into a format suitable for analysis. This may involve converting the data into a standard format such as GeoJSON or shapefile. The next step is to preprocess the data, which may involve cleaning the data, removing duplicates, and handling missing or invalid values. The preprocessed geospatial data is typically represented as a dataframe with columns containing the location coordinates, attributes, and other metadata.

Once the data is preprocessed, the next step is to visualize the data to gain insights and identify patterns. This may involve creating maps, heatmaps, and other types of visualizations using libraries such as matplotlib, seaborn, and folium. Visualizing geospatial data allows analysts to identify patterns and trends, such as areas with high or low density of certain features, and make informed decisions based on those insights.

Another important step in working with geospatial data is spatial analysis, which involves analyzing the relationships between spatial features and attributes. This may involve performing spatial queries such as selecting features within a certain distance or area, or aggregating features based on their attributes. Spatial analysis can be performed using libraries such as GeoPandas, Shapely, and PySAL.

Geospatial data analysis can be used in a variety of applications such as urban planning, transportation, ecology, and geology. For example, urban planners can use geospatial data to analyze traffic patterns and plan new transportation routes. Ecologists can use geospatial data to study the distribution of plant and animal species and their habitats. Geologists can use geospatial data to study the formation and distribution of geological features such as rocks and minerals.

Pandas for Geospatial Analysis

Support for GeoDataFrame
Pandas 2.0 provides support for GeoDataFrame, which is an extension of the DataFrame object that includes geometry columns. Geometry columns can store various types of spatial data such as points, lines, and polygons. The GeoDataFrame object allows users to perform spatial operations and analysis on the data using various Python libraries such as Shapely and GeoPandas.

Integration with other Python geospatial libraries
Pandas 2.0 can be easily integrated with other Python geospatial libraries such as GeoPandas, Shapely, and folium. These libraries provide additional functionality for geospatial analysis such as spatial joins, buffer analysis, and interactive map visualizations. By using Pandas 2.0 in conjunction with these libraries, users can perform complex geospatial analysis tasks efficiently and effectively.

Flexible data manipulation capabilities
Pandas 2.0 provides powerful data manipulation capabilities such as filtering, aggregation, and pivoting, which can be applied to geospatial data as well. These operations can be used to extract insights and patterns from the geospatial data and to prepare the data for further analysis and visualization.

Support for time series analysis
Pandas 2.0 has built-in support for time series analysis, which can be useful for analyzing and visualizing geospatial data over time. For example, users can aggregate the geospatial data by time intervals and create time series plots to visualize trends and patterns over time.

Overall, Pandas 2.0 provides a powerful and flexible framework for geospatial data analysis. By leveraging its support for GeoDataFrame, integration with other Python geospatial libraries, data manipulation capabilities, and time series analysis, users can perform complex geospatial analysis tasks efficiently and effectively.

Working with Geospatial Data Formats

Pandas 2.0 provides support for several geospatial data formats that are commonly used in data analysis:

GeoJSON

GeoJSON is a popular geospatial data format for encoding various types of geometries such as points, lines, and polygons. It is a lightweight format that is easy to use and read. Pandas 2.0 supports reading and writing GeoJSON data using the read_json() and to_json()

methods, respectively.

ESRI Shapefile

ESRI Shapefile is a widely used geospatial data format for storing various types of spatial data such as points, lines, and polygons. It consists of several files that together represent the geometry and attributes of the spatial features. Pandas 2.0 supports reading and writing Shapefile data using the read_file() and to_file() methods, respectively.

GPS Exchange Format (GPX)

GPX is a standard file format for exchanging GPS data between various applications and devices. It consists of an XML-based format that stores various types of GPS data such as tracks, routes, and waypoints. Pandas 2.0 supports reading GPX data using the read_xml() method from the xml.etree.ElementTree module.

Keyhole Markup Language (KML)

KML is a file format for storing various types of geospatial data such as points, lines, and polygons. It is an XML-based format that is commonly used for creating maps and visualizations. Pandas 2.0 supports reading KML data using the read_xml() method from the xml.etree.ElementTree module.

GeoTIFF

GeoTIFF is a raster geospatial data format that is widely used for storing various types of spatial data such as satellite imagery and terrain models. It is a TIFF-based format that includes additional spatial metadata such as coordinate reference system and georeferencing information. Pandas 2.0 supports reading GeoTIFF data using the gdal library.

Overall, Pandas 2.0 provides support for several geospatial data formats that are commonly used in data analysis. By leveraging its support for these formats, users can easily load and preprocess geospatial data in various formats and perform analysis and visualization using other Python libraries such as Shapely and GeoPandas.

Load, Explore, Transform, and Save Geospatial Data

Given below are the steps to load, explore, transform, and save geospatial data from the IndiaGIS dataset:

Download and Extract Dataset

The IndiaGIS dataset is available as a compressed ZIP file. Download the file from the link provided and extract its contents to a local directory.

Load Data using GeoPandas

GeoPandas is a Python library that extends the functionality of Pandas to support geospatial data. Install GeoPandas using pip install geopandas if it is not already installed. Then, load the dataset using the read_file() method from GeoPandas:

```python
import geopandas as gpd


# load the data
india_gdf = gpd.read_file('path/to/indiaGIS.shp')
```

This will load the shapefile data into a GeoDataFrame object named india_gdf.

Explore the Data

Once the data is loaded, you can explore its attributes and geometry using various methods and attributes of the GeoDataFrame object. For example, you can use the head() method to view the first few rows of the data:

```python
print(india_gdf.head())
```

You can also use the plot() method to create a basic visualization of the data:

```python
india_gdf.plot()
```

This will create a map of India with the geometries of the features in the dataset.

Transform the Data

You can perform various transformations on the data using GeoPandas methods and functions. For example, you can use the to_crs() method to transform the coordinate reference system (CRS) of the data to a different CRS:

```python
# transform the CRS to EPSG:4326
india_gdf = india_gdf.to_crs(epsg=4326)
```

This will transform the CRS of the data to the commonly used WGS84 CRS with EPSG code 4326.

Save the Data

You can save the data in a new format using the to_file() method of the GeoDataFrame object. For example, you can save the data as a GeoJSON file using the following code:

```python
# save the data as a GeoJSON file
india_gdf.to_file('path/to/indiaGIS.geojson', driver='GeoJSON')
```

This will save the data as a GeoJSON file in the specified directory.

Overall, the above steps demonstrate how to load, explore, transform, and save geospatial data using GeoPandas. GeoPandas provides a powerful and flexible framework for geospatial data analysis and can be used in conjunction with other Python libraries such as Shapely, folium, and matplotlib for advanced analysis and visualization.

Advanced Geospatial Manipulation

Following are some more advanced geospatial manipulation and data transformation techniques using GeoPandas:

Spatial Joins

Spatial joins are an important technique used in geospatial analysis that helps to combine two or more datasets based on their location and spatial relationships. In essence, spatial joins enable you to overlay geospatial data to identify patterns, relationships, and trends that may not be immediately apparent when the datasets are considered in isolation.

One of the most common types of spatial join is the point-in-polygon join. This type of join involves combining a point dataset with a polygon dataset to determine which points fall within each polygon. This can be useful for analyzing the distribution of events, such as crime incidents, within specific geographic areas or regions. Other types of spatial joins include polygon-on-polygon, line-on-polygon, and point-on-line joins. These joins allow you to combine datasets based on other spatial relationships, such as determining which polygons intersect or which lines intersect with polygons.

Spatial joins are commonly used in a wide range of applications, including urban planning, natural resource management, and environmental monitoring. By combining geospatial datasets, you can gain a deeper understanding of the relationships between different variables and make more informed decisions about how to manage and allocate resources in a given area.

Performing Spatial Join Operations

To perform spatial joins on the given database, you'll need a geospatial library like GeoPandas, which extends the functionality of Pandas to handle geospatial data. In this example, we'll assume you have two datasets - one from the indiaGIS.zip file and another dataset containing some geospatial information. We will use Python and the GeoPandas library for this task.

First, install the required libraries:

```
pip install geopandas
```

Import the necessary libraries:

```
import pandas as pd
import geopandas as gpd
from shapely.geometry import Point
```

Load the indiaGIS dataset as a GeoDataFrame:

```
# Assuming the data is in a CSV format with 'latitude' and 'longitude' columns
data = pd.read_csv('path/to/your/indiaGIS.csv')


# Convert the DataFrame to a GeoDataFrame by creating a 'geometry' column
geometry = [Point(xy) for xy in zip(data.longitude, data.latitude)]
data_gdf = gpd.GeoDataFrame(data, geometry=geometry)
```

Load the second dataset as a GeoDataFrame. Replace 'your_second_dataset.shp' with the appropriate file path and format (e.g., Shapefile, GeoJSON).

```python
second_dataset_gdf = gpd.read_file('path/to/your/second_dataset.shp')
```

Perform the spatial join:

```python
# You can choose 'inner', 'left', or 'right' as the type of join
joined_gdf = gpd.sjoin(data_gdf, second_dataset_gdf, how='inner',
op='intersects')
```

In the gpd.sjoin() function, the how parameter defines the type of join (inner, left, or right), and the op parameter defines the type of spatial operation used for the join (intersects, within, or contains).

Save the result of the spatial join:

```python
joined_gdf.to_file('path/to/your/joined_data.shp')
```

This example demonstrates how to perform a spatial join using the GeoPandas library in Python. Make sure to replace the file paths and formats with the appropriate values for your datasets.

Buffer Analysis

Buffer analysis is a common technique used in Geographic Information Systems (GIS) to analyze and visualize spatial data. It involves creating a buffer zone around a particular feature or set of features, such as roads, rivers, buildings, or points of interest. The buffer zone represents a specific distance or radius around the feature, and can be used to identify areas that fall within this distance.

Buffer analysis is particularly useful in urban planning, environmental studies, and emergency management. For example, urban planners can use buffer analysis to determine the impact of proposed new developments on existing infrastructure. Environmental scientists can use buffer analysis to identify areas of critical habitat or to model the spread of pollutants. Emergency managers can use buffer analysis to identify areas at risk of flooding or other natural disasters. GIS software, such as ArcGIS and QGIS, offer a range of tools for performing buffer analysis. These tools allow users to specify the feature or features to buffer, the buffer distance, and the type of buffer (e.g., circular, square, or custom polygon). The resulting buffer zone can then be used for further analysis, such as overlaying with other datasets or generating statistics.

Performing Buffer Analysis

To perform buffer analysis on the given database, we will use Python and the GeoPandas library.

Install the required libraries if you haven't already:

```
pip install geopandas
```

Import the necessary libraries:

```python
import pandas as pd
import geopandas as gpd
from shapely.geometry import Point
```

Load the indiaGIS dataset as a GeoDataFrame:

```python
# Assuming the data is in a CSV format with 'latitude' and 'longitude' columns
data = pd.read_csv('path/to/your/indiaGIS.csv')

# Convert the DataFrame to a GeoDataFrame by creating a 'geometry' column
geometry = [Point(xy) for xy in zip(data.longitude, data.latitude)]
data_gdf = gpd.GeoDataFrame(data, geometry=geometry)
```

Define the buffer distance:

```python
buffer_distance = 1000  # meters, or any other unit depending on your dataset's CRS
```

Perform buffer analysis:

```python
# Make sure your data is in a projected coordinate system with units in meters
# If not, you need to reproject it to an appropriate CRS before buffering
data_gdf_projected = data_gdf.to_crs("EPSG:32643")  # Example: UTM Zone 43N (WGS 84)
```

```
# Create the buffer polygons around the points
buffered_gdf = data_gdf_projected.copy()
buffered_gdf['geometry'] = data_gdf_projected.buffer(buffer_distance)
```

If you want to perform further spatial analysis, reproject the buffered dataset back to the original CRS:

```
buffered_gdf_original_crs = buffered_gdf.to_crs(data_gdf.crs)
```

Save the buffer polygons as a new geospatial file:

```
buffered_gdf_original_crs.to_file('path/to/your/indiaGIS_buffered.shp')
```

This example demonstrates how to perform buffer analysis using the GeoPandas library in Python. Make sure to replace the file paths and formats with the appropriate values for your datasets, and adjust the buffer distance and CRS as needed.

Dissolve

Dissolve is a crucial geospatial operation that is widely used in geographic information systems (GIS) and geospatial data analysis. It involves combining features within a dataset that share a common attribute or characteristic, such as a region or a city. The process merges the geometries of these features and aggregates their attributes, resulting in a single feature with a new set of attributes.

The dissolve operation can be used for various purposes, such as simplifying complex datasets, creating summary statistics, or generating new spatial data layers. For instance, if you have a map of the United States showing state boundaries and population data, you can use dissolve to merge all the states with a population below a certain threshold into a single region. This process simplifies the data and makes it easier to analyze.

Performing Dissolve Operation

To perform a dissolve operation on the given database, we will use Python and the GeoPandas library.

Install the required libraries if you haven't already:

```
pip install geopandas
```

Import the necessary libraries:

```python
import pandas as pd
import geopandas as gpd
from shapely.geometry import Point
```

Load the indiaGIS dataset as a GeoDataFrame:

```python
# Assuming the data is in a CSV format with 'latitude' and 'longitude' columns
data = pd.read_csv('path/to/your/indiaGIS.csv')

# Convert the DataFrame to a GeoDataFrame by creating a 'geometry' column
geometry = [Point(xy) for xy in zip(data.longitude, data.latitude)]
data_gdf = gpd.GeoDataFrame(data, geometry=geometry)
```

Choose an attribute for the dissolve operation. In the given example, we'll use the 'group_column' attribute. Replace 'group_column' with the appropriate column name in your dataset:

```python
dissolve_column = 'group_column'
```

Perform the dissolve operation:

```python
dissolved_gdf = data_gdf.dissolve(by=dissolve_column)
```

By default, the dissolve operation will aggregate the other attributes by taking their first value in the group. If you need to apply different aggregation functions, you can use the aggfunc parameter:

```python
dissolved_gdf = data_gdf.dissolve(by=dissolve_column, aggfunc={'column1': 'sum', 'column2': 'mean'})
```

Replace 'column1' and 'column2' with the appropriate column names in your dataset and choose the aggregation functions (e.g., 'sum', 'mean', 'max', 'min') that best suit your needs.

Save the dissolved dataset as a new geospatial file:

```
dissolved_gdf.to_file('path/to/your/indiaGIS_dissolved.shp')
```

This example demonstrates how to perform a dissolve operation using the GeoPandas library in Python. Make sure to replace the file paths, formats, and column names with the appropriate values for your datasets.

Overlay Analysis

Overlay analysis is a process that allows geospatial data to be combined and analyzed by performing geometric set operations, such as union, intersection, or difference. This technique is commonly used in GIS (Geographic Information Systems) to extract relevant information from different datasets and obtain new insights into the relationships between them.

To perform overlay analysis, we can use Python and the GeoPandas library, which provides a set of tools and functions for working with geospatial data. With GeoPandas, we can load and manipulate geospatial datasets, perform various spatial operations, and visualize the results using matplotlib or other plotting libraries. For example, if we have a dataset of buildings and a dataset of land parcels, we can use overlay analysis to determine which buildings are located within each parcel or which parcels are adjacent to specific buildings. This information can be useful for various applications, such as urban planning, environmental analysis, or real estate management.

Performing Overlay Analysis

Assuming you have two datasets - one from the indiaGIS.zip file and another dataset containing some geospatial information, follow these steps:

Install the required libraries if you haven't already:

```
pip install geopandas
```

Import the necessary libraries:

```python
import pandas as pd
import geopandas as gpd
from shapely.geometry import Point
```

Load the indiaGIS dataset as a GeoDataFrame:

```python
# Assuming the data is in a CSV format with 'latitude' and 'longitude' columns
data = pd.read_csv('path/to/your/indiaGIS.csv')

# Convert the DataFrame to a GeoDataFrame by creating a 'geometry' column
geometry = [Point(xy) for xy in zip(data.longitude, data.latitude)]
data_gdf = gpd.GeoDataFrame(data, geometry=geometry)
```

Load the second dataset as a GeoDataFrame. Replace 'your_second_dataset.shp' with the appropriate file path and format (e.g., Shapefile, GeoJSON).

```python
second_dataset_gdf = gpd.read_file('path/to/your/second_dataset.shp')
```

Ensure both datasets have the same CRS (Coordinate Reference System). If not, reproject one of the datasets to match the other:

```python
second_dataset_gdf = second_dataset_gdf.to_crs(data_gdf.crs)
```

Perform the overlay analysis. In this example, we'll perform an intersection operation:
python

```python
overlay_gdf = gpd.overlay(data_gdf, second_dataset_gdf, how='intersection')
```

You can change the how parameter to perform other set operations, such as 'union' or 'difference'.

Save the result of the overlay analysis:

```python
overlay_gdf.to_file('path/to/your/indiaGIS_overlay.shp')
```

This example demonstrates how to perform overlay analysis using the GeoPandas library in Python. Make sure to replace the file paths, formats, and column names with the appropriate values for your datasets.

Overall, these advanced geospatial manipulation and data transformation techniques using GeoPandas can help to perform complex analysis and modeling tasks on geospatial data.

Geocoding and Reverse Geocoding

Geocoding and reverse geocoding are techniques used to convert geographic data between different formats, specifically between human-readable addresses and geographic coordinates.

Geocoding

Geocoding is the process of translating a location description, such as a street address, postal code, or place name, into geographic coordinates. It involves identifying the spatial location of a particular address or place on the earth's surface by converting it into latitude and longitude values. Geocoding is a vital tool used in a wide range of applications, including logistics, mapping, and navigation systems, social media platforms, and web-based services. Geocoding enables a user to locate a specific point on the map, make informed decisions based on location data, and analyze spatial patterns.

Geocoding uses a combination of geospatial data and algorithms to accurately determine the geographic coordinates of a particular location. The process of geocoding involves several steps, starting with data acquisition, normalization, and matching. Data acquisition involves collecting data sources that contain location information, such as postal code directories, street maps, and satellite imagery. Normalization is the process of standardizing the location information by correcting spelling mistakes, abbreviations, and other inconsistencies. Matching involves comparing the normalized data with a geographic reference database to identify the latitude and longitude of the location accurately.

Reverse Geocoding

Reverse geocoding, on the other hand, is the process of translating geographic coordinates into a human-readable address or place name. Reverse geocoding is useful in situations where a user has the latitude and longitude values of a location and needs to know its physical address. Reverse geocoding is also used in applications where users want to obtain location-specific information, such as weather forecasts, nearby restaurants, or tourist attractions.

Reverse geocoding involves a similar process as geocoding, but in reverse. The process starts with acquiring the geographic coordinates of a location and then matching them with a geographic reference database to determine the corresponding address or place name. Reverse geocoding typically uses the same geospatial data and algorithms as geocoding, but in reverse. The accuracy of reverse geocoding depends on the quality of the geospatial data used and the matching algorithm employed.

Implement Geocoding and Reverse Geocoding

Geocoding and reverse geocoding have numerous applications in various industries, including transportation, logistics, marketing, and real estate. In transportation, geocoding and reverse geocoding are used for route optimization, fleet management, and vehicle tracking. In logistics, geocoding and reverse geocoding are used for warehouse location optimization, shipment tracking, and delivery routing. In marketing, geocoding and reverse geocoding are used for targeted advertising and location-based promotions. In real estate, geocoding and reverse geocoding are used for property valuation, site selection, and market analysis.

To implement geocoding and reverse geocoding on the given database, follow these steps:

Download and extract the indiaGIS.zip file from the provided link.

The extracted data should be in a geospatial file format, such as GeoJSON, Shapefile, or CSV with latitude and longitude columns.

Choose a programming language and a library to perform geocoding and reverse geocoding. For this example, we'll use Python and the 'geopy' library.

Install the required library:

```
pip install geopy
```

Import the necessary modules and read the geospatial data:

```
import pandas as pd
from geopy.geocoders import Nominatim
from geopy.extra.rate_limiter import RateLimiter


# Assuming the data is in a CSV format with 'latitude' and 'longitude' columns
```

```python
data = pd.read_csv('path/to/your/indiaGIS.csv')
```

Create a geolocator instance and rate limiter:

```python
geolocator = Nominatim(user_agent="your_app_name")
rate_limiter = RateLimiter(geolocator.geocode, min_delay_seconds=1)
```

Implement geocoding:

```python
def geocode_address(address):
    location = rate_limiter(address)
    if location:
        return location.latitude, location.longitude
    else:
        return None, None

data['coordinates'] = data['address'].apply(geocode_address)
data[['latitude', 'longitude']] = pd.DataFrame(data['coordinates'].tolist(),
index=data.index)
```

Replace 'address' with the appropriate column name in your dataset containing addresses.

Implement reverse geocoding:

```python
def reverse_geocode(coordinates):
    location = geolocator.reverse(coordinates, exactly_one=True)
    if location:
        return location.address
    else:
        return None

data['reverse_geocoded_address'] = data['coordinates'].apply(reverse_geocode)
```

Save the updated data:

```
data.to_csv('path/to/your/indiaGIS_updated.csv', index=False)
```

This example demonstrates how to use the geopy library to perform geocoding and reverse geocoding on a dataset. Please note that the geopy library uses OpenStreetMap's Nominatim service for geocoding, which has usage limits. Make sure to check their usage policy and respect it. You can also use other geocoding services such as Google Maps API, Bing Maps API, or Mapbox, but these may require API keys and have different usage limits.

Summary

This chapter discussed various aspects of geospatial data analysis, including data formats, the role of Pandas, and several geospatial operations using Python and GeoPandas.

Geospatial data formats store information about the location and shape of geographic features, along with their attributes. Common formats include Shapefiles, GeoJSON, and CSV files with latitude and longitude columns. These formats enable the storage, visualization, and manipulation of spatial data in a variety of applications.

Although not specifically designed for geospatial data, Pandas can contribute to geospatial analysis by providing efficient data structures like DataFrames to manage tabular data, including geospatial attributes. However, for more advanced geospatial operations, GeoPandas, which extends Pandas functionality, is recommended.

The chapter covered a series of geospatial operations using Python and GeoPandas:

Spatial Joins: Spatial joins combine two datasets based on the spatial relationship between their geometries. The chapter described how to perform spatial joins using the GeoPandas library by loading datasets as GeoDataFrames, and using the sjoin function.

Buffer Analysis: Buffer analysis creates zones around specific features in a dataset, based on a specified distance. The chapter illustrated how to perform buffer analysis using GeoPandas by converting a DataFrame to a GeoDataFrame and using the buffer function.

Dissolve Operation: Dissolve is a geospatial operation used to combine features based on a common attribute, merging their geometries and aggregating their attributes. The chapter showed how to perform a dissolve operation using GeoPandas, by loading a dataset as a GeoDataFrame and using the dissolve function.

Overlay Analysis: Overlay analysis combines two or more datasets by performing set operations on their geometries, such as intersection, union, or difference. The chapter demonstrated how to perform overlay analysis using GeoPandas, by loading datasets as GeoDataFrames, ensuring they have the same CRS, and using the overlay function.

Geocoding and Reverse Geocoding: Geocoding is the process of converting addresses or place names into geographic coordinates (latitude and longitude), while reverse geocoding is the opposite. The chapter demonstrated how to implement these processes using the 'geopy' library.

Throughout the chapter, we provided examples using Python, Pandas, and GeoPandas libraries to perform these geospatial operations on the given dataset. This knowledge enables users to manipulate and analyze geospatial data effectively, making it suitable for a wide range of applications in the geospatial domain.

Thank You

Epilogue

As you close the pages of "Learning Pandas 2.0," you can't help but feel empowered by the depth and breadth of knowledge you've acquired. This comprehensive guide has taken you on a journey through the inner workings of the latest version of the popular Python library for data manipulation and analysis. You've learned how to unlock the full potential of Pandas 2.0 and its new features, and gained valuable insights and hands-on experience to ensure you're equipped with the knowledge and skills to master it.

You started with the core concepts of Pandas 2.0, from data structures to advanced indexing techniques, and explored the myriad ways you can load, store, and manipulate data. You've gained expertise in time series and DateTime operations, and visualized your data using powerful tools like Seaborn and Plotly. You've also learned how to improve the performance of your Pandas code through optimization strategies and parallel computing with Dask.

As you progressed, you've also learned how to unleash the power of machine learning by integrating Pandas 2.0 with Scikit-learn, TensorFlow, and PyTorch. You've mastered techniques for handling imbalanced data and feature scaling, and delved into the world of text data and natural language processing with Pandas 2.0.

Moreover, you've expanded your knowledge with geospatial data analysis and visualization, and applied the skills you've learned through real-world case studies in finance, social media, e-commerce, IoT, and healthcare. These hands-on exercises helped you gain valuable experience in applying Pandas 2.0 to real-world scenarios, preparing you to tackle complex data analysis tasks in your professional life.

Finally, the book also provided insights into best practices and future trends in Pandas 2.0, setting you on the path to becoming a Pandas expert. You feel confident in your ability to harness the power of this cutting-edge library, and excited to apply your newfound knowledge to your future data analysis projects.

As you put down the book, you can't help but be grateful for the all-encompassing guide that has been "Learning Pandas 2.0." Whether you're a seasoned data professional or just starting your journey in data science, this book has been the perfect resource to help you elevate your data analysis skills. With its step-by-step instructions, real-world examples, and valuable insights, you are now fully equipped to master Pandas 2.0 and unlock its full potential.

Made in the USA
Monee, IL
07 July 2026